# WATER WAVES

# THE WYKEHAM SCIENCE SERIES
**for schools and universities**

*General Editors:*
Professor Sir Nevill Mott, F.R.S.
Cavendish Professor of Physics
University of Cambridge

G. R. Noakes
Formerly Senior Physics Master
Uppingham School

To broaden the outlook of the senior grammar school pupil and to introduce the undergraduate to the present state of science as a university study is the aim of the Wykeham Science Series. Each book seeks to reinforce this link between school and university levels, and the main author, a university teacher distinguished in the field, is assisted by an experienced sixth-form schoolmaster.

# WATER WAVES

N. F. Barber – University of Wellington

*Assisted by*
G. Ghey

**WYKEHAM PUBLICATIONS (LONDON) LTD**
(A subsidiary of Taylor & Francis Ltd)
**LONDON & WINCHESTER**
**1969**

First published 1969 by Wykeham Publications (London) Ltd.

© 1969 N. F. Barber. All rights reserved. No part of this publication may be reproduced, stored in a retrieval system, or transmitted, in any form or by any means, electronic, mechanical, photocopying, recording, or otherwise, without the prior permission of the copyright owner.

Cover illustration by courtesy of Keystone Press Agency Ltd.

85109 060 5

Printed in Great Britain by Taylor & Francis Ltd.
Cannon House, Macklin Street, London, W.C.2

TC
172.
B 36

World wide distributon, excluding the Western Hemisphere, India, Pakistan and Japan, by Associated Book Publishers, Ltd., London and Andover.

# PREFACE

THE occasional numerical illustrations in the book are in SI units except where it is more convenient to use knots and sea miles so that distances relate in a simple way to degrees of latitude on a chart. In accordance with what seems to be the present convention, SI units are not given an ' s ' in the plural, so that the text speaks of a wave having a period of 20 second, not 20 seconds, and a wavelength of 300 metre, not 300 metres. The ' s ' has been retained however in speaking of minutes, miles, hours and degrees of arc, which are not units in the SI system.

The mathematical arguments, which are mainly confined to chapters 2 and 3, assume an acquaintance with calculus to the extent of knowing the differential coefficients of the sine and cosine and of powers of the variable. Exponential functions are also used but an explanation is given in the text. Some of the formulae should properly be written with the $\partial$ symbol for partial differentiation (as explained in a footnote) but the ordinary d has been retained in an attempt to avoid giving too strange an appearance to the mathematics.

*Wellington*
*July* 1969                                                    N. F. BARBER

# LIST OF SYMBOLS

The symbols used in the text for physical quantities are as follows :

$A$ and $B$    for vertical and horizontal amplitudes of particle motion in a wave (chap. 1)
$A$    for a cross-sectional area (chap. 2, 3)
$a$ and $b$    for amplitudes in two harmonic motions (chap. 4)
$a$    for acceleration (chap. 1, 2)
$c$    for wave speed
$D$ (and $h$)    for depth in water
$g$    for acceleration due to gravity
$k$    for wave-number or $2\pi/$(wavelength)
$L$ or $l$    for a length or distance
$M$ or $m$    for a mass
$p$    for pressure
$q$    for the speed of a particle where $q^2 = u^2 + v^2 + w^2$
$r$    for a radial distance
$t$    for time
$u, v, w$    for rectangular components of particle velocity
$x, y, z$    for rectangular components of position

The greek symbols used in the text are :

$\alpha$ (alpha)    for an angle in radian measure
$\Delta$ or $\delta$ (delta)    to denote a small change or difference
$\theta$ (theta)    for an angle
$\pi$ (pi)    for 3.14...
$\phi$ (phi)    for the velocity potential explained in the text

# CONTENTS

*Preface*     v

*Chapter 1*     LOOKING AT WAVES     1
    The wave is not the water—acceleration and the false vertical—seeking the simplest idea—guessing the motion in a wave—the speed of waves—tides and tsunamis

*Chapter 2*     THE MOTION OF A FRICTIONLESS FLUID     20
    Physics deals with an 'idealized' world—the ideal fluid—the circulation theorem—waves are an irrotational motion—the velocity potential—how to represent an incompressible fluid in mathematics

*Chapter 3*     THE SPEED OF A WAVE-TRAIN     36
    Inventing a 'steady state'—guessing a formula—the separation of variables—a formula for pressure—testing the formula—a formula for waves over a flat sea-bed—the end of the mathematics

*Chapter 4*     MAKING UP WAVE PATTERNS     56
    Waves going the same way—waves going opposite ways—waves crossing—waves between four walls—seiches in lakes—rotating waves—waves in a rectangular basin—circular waves—ripples, or surface tension waves

*Chapter 5*     GROUPS OF WAVES     76
    The curious behaviour of a group of waves—a formula for the group velocity—the travel of swell in the North Atlantic—swell travelling from Cape Horn to Cornwall—swell moving through tidal streams—swell travelling from Australia to California—the energy in waves—observing waves at sea—waves inside the sea (internal waves)—waves coming ashore—refraction of waves

*Chapter 6*　　MORE ABOUT SEA WAVES　　102
　　How sea waves can shake the ocean bed—detecting microseisms—the beat of the surf—inventing the idea of a correlation coefficient—measuring correlations—the surf beat and the surges

*Chapter 7*　　THE WAKE OF A SHIP　　118
　　How waves spread from a centre—each wave crest accelerates—constructing the pattern of the wake of a ship—the critical speed in shallow water—the wave-making resistance to a ship—ripple wakes are ahead, not behind—the smooth track left by a ship—waves riding on streams—the wind and the tide—can waves emerge from the place where they break?—goodbye

*A Postscript*　　140

*Subject Index*　　141

*The Wykeham Series*　　143
　　Science and Technological

# CHAPTER 1
## looking at waves

SOME people swim through breaking waves. They dive down under each wave and swim through as it passes over them.

Other people race boats through the surf, though one would scarcely think it possible.

Fig. 1.1. People doing something with waves.

Others use surf-boards to ride the waves. People feel an urge to do something with waves.

Some people like to argue about waves, and this is a book for people who like to argue about them.

### What is a wave?

If you watch the sea from the shore you can sometimes trace a wave all the way as it comes in from the open sea. First a long line of shadow scudding towards the land, then an obvious ridge of water growing higher as it advances, then a foaming top and a curling breaker if the beach is steep.

The water that forms the top of a curling breaker has spun upward and forwards and is on the point of dropping. Was this water carried along by the wave from the open sea? The answer is 'no'. All the water in the breaking wave is water that was sitting at the beach long before the wave arrived. The wave is not the water.

Fig. 1.2. His grandfather's axe.

So it seems that when you watch a wave approaching from the sea, you are looking every few seconds at quite a different piece of water. This is rather like the story of the man who was proud of an axe that belonged to his grandfather, and said it had needed only two new heads and three new handles since his grandfather's time. The axe he showed was made of material that his grandfather never saw. What right had he to call it his grandfather's axe?

Yet it seems quite sensible to say that something travelled across the sea to the shore. If it wasn't the water that travelled, what was it? I think one could say it was a shape, a pattern of outline and water motion.

In this book I shall use the word 'wave' to mean the shape, the pattern. It can advance across the water and one can set a number to the speed at which it travels and call it the 'wave speed'. Remember that this is not the speed of travel of the water itself; any particular piece of water moves up and down, forward and backward as the wave passes it. When I want to speak of the motion of the water itself I will call it the 'water velocity' or the 'particle velocity'.

The water is of course a more permanent and 'real' thing than the wave. Water can move but it must go somewhere, it can't disappear. But a wave, which is merely a pattern, can change. It may become more marked or it may fade out and disappear. Indeed this usually happens to waves. When a wave doesn't disappear in the course of a few seconds but can be followed for a long way, it strikes one as unusual. If you like to think so, it is interesting. One wonders why it doesn't disappear.

*The tilted surface*

The front of a breaking wave can look quite smooth and glassy, like the surface of water at rest in a bucket. But the front of the breaking wave is tilted quite steeply.

You know that with care you can swing a bucket of water, horizontally in a circle, so that the bucket tilts quite steeply, yet you need spill no water. This is because the surface of the water tilts too. Is something like this happening in the wave?

Fig. 1.3. You need spill none.

A water surface sets itself at every instant at right angles to what some people call the 'false vertical'. I must explain what is meant by this.

You can find the true vertical by hanging something on a long string from a fixed support. The object on the end is not being accelerated, so the forces acting on the object must add up to zero. There are only two, the force of gravity and the pull of the string, and to balance they must act in opposite directions. The line of the string therefore shows the direction of gravity, the vertical†.

You could do the same thing in a smoothly moving train or motor car and still get the true vertical so long as the carriage was moving at a constant velocity. When there is no acceleration the pull of the string must just balance the force of gravity, as before.

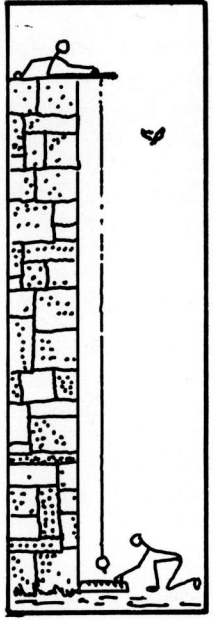

Fig. 1.4. The true vertical.

But if the carriage or car is accelerating the string deflects away from the true vertical. You would see this if the car travelled round a curve in the road, or if it were increasing speed along a straight road, or if it were gathering speed as it went down a hill. In each case the object hanging on the string needs to be accelerated in order to keep pace with the car. When the acceleration first begins, the hanging object tends to be left behind and the string deflects from the true vertical. When things have become steady, the pull of the inclined string and the force of gravity combine to give just sufficient force to give the object the same acceleration as the carriage.

You could calculate the direction of the string even before doing the experiment if you knew what the acceleration would be. The

† A builder does this to get his walls vertical. He uses something heavy but small so that wind disturbs it little, perhaps a lead weight. He calls it a 'plumb line' because *plumbum* is the Latin word for lead.

pull of the string would need to be equivalent to two forces, *mg* vertically upwards to oppose gravity and *ma* in the direction of the acceleration (where *a* is the magnitude of the acceleration). So you find the resultant by a parallelogram construction and the direction of this resultant shows the line along which the string lies. In fig. 1.5

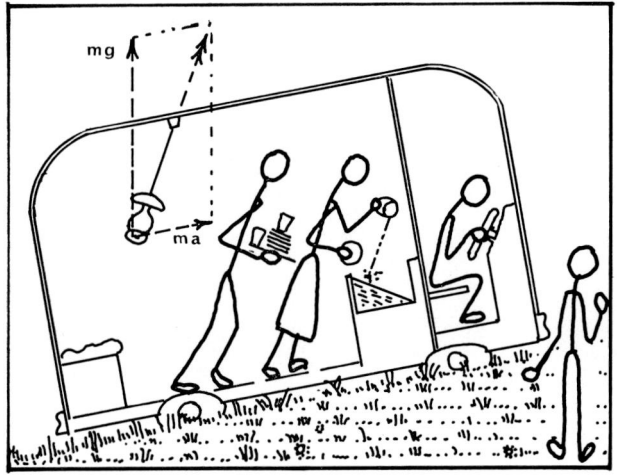

Fig. 1.5. The false vertical.

I have drawn the acceleration somewhat upwards from the horizontal; perhaps the car is accelerating as it climbs a hill.

The line of the string when the carriage accelerates is not the true vertical and yet it shows a kind of 'vertical'. If you wanted to stand in the accelerating carriage without holding on to anything you would need to lean until you were standing parallel to the string. If you held something and dropped it, the direction in which it would seem to you to fall would be parallel to the line of the string. If you had a bucket of water in the carriage the surface of the water would tilt and set itself at right angles to the direction of the string. So some people call this direction the 'false vertical'.

Why does the water surface tilt? Well, why doesn't the water fall through the bottom and sides of the bucket? Because the sides and bottom press on the water. But what about a piece of water inside the rest, why does that stay where it is? Because the surrounding water presses on it.

Imagine a rectangular-shaped piece of water inside all the rest. I am going to choose it so that the sides are parallel or perpendicular to the false vertical. This is because I know that the total thrust on this block of water needs to be in the direction of the false vertical.

Then, like the object on the string, the force of gravity can be supported, and at the same time there is a further force just sufficient to accelerate it so that it keeps pace with the carriage.

Now, you know that the thrust of water acts at right angles to its boundary. The thrusts on the 'top' and 'bottom' of the rectangular block are in the direction at right angles to them. They are parallel

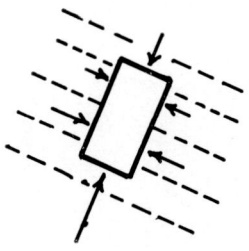

Fig. 1.6. A piece of water inside the rest.

to the false vertical and no doubt the difference of thrust at top and bottom will adjust itself so that the total will be just right to keep the water in place in the bucket. The thrust will be greater on the 'bottom' than on the 'top'.

But what of the thrusts on the sides of the block? These thrusts are at right angles to the false vertical. They must exactly cancel one another. The areas of opposite sides are equal so this suggests that the pressure (thrust per unit area) is the same on all sides. This is too simple a view of course, because we have seen that the pressure is different at the top and at the bottom of the block. Perhaps the following is a better picture of the situation: if one draws any line (or more correctly a plane surface) at right angles to the false vertical, then it seems that at all points on this surface the water pressure will be the same. The broken lines in fig. 1.6 illustrate a number of these 'contours of constant pressure'. These surfaces of constant pressure are at right angles to the false vertical; they must be so if the water is to accelerate with the bucket.

Then as a final question, do we expect the free surface of the water to be a surface of constant pressure? If it is, then our argument says that it should be at right angles to the false vertical.

If there were no air over the water then the pressure on the free water surface would be zero. This would make the pressure constant (zero). But there is air above the water. If the air is moving you may have different pressures at different places. For example, if you blow down through a tube on a water surface you can make a dimple in the surface, suggesting that where the air is brought to rest, just below the tube, the pressure is greater than at the sides, where the air is moving outwards. I shall suppose, however, that the air over the bucket in the carriage is all moving with it, is all at rest relative to the carriage. Then the air pressure on the water will be sensibly the same everywhere. The free surface of the water will be a surface of constant pressure and it will be at right angles to the false vertical.

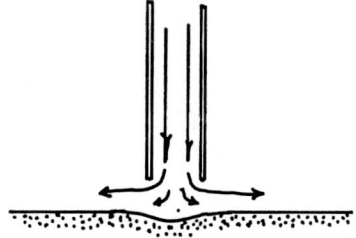

Fig. 1.7. The pressure is lower where the air moves than where it is brought to rest.

Now we can think of the wave again. When the water surface tilts in a wave, it seems that this must come about because the water is being accelerated. In fig. 1.8 I have sketched part of a wave outline and I am thinking of the water near the point O. The line OP at right angles to the surface near point O must be the false vertical for that piece of water.

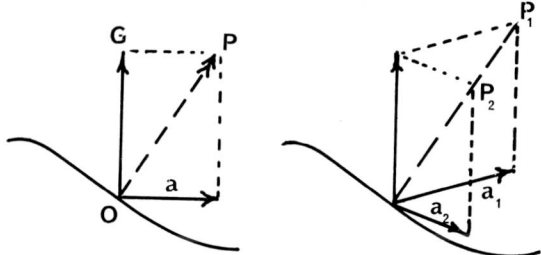

Fig. 1.8. A tilted surface shows that the water is being accelerated.

One can then estimate how the water at point O is accelerating. In fig. 1.8 I have drawn the true vertical OG and drawn the arrow proportional to $g$, the acceleration due to gravity, but drawn it upwards. When this is combined with the arrow (vector) representing the acceleration $a$ of the water, the resultant must lie along the line OP. The same construction was used in discussing the object hanging on a string in fig. 1.5. If I could assume that the acceleration were horizontal, then the construction would show me just how large the acceleration was. I cannot assume this; the acceleration might be somewhat upwards or somewhat downwards, and these two cases are sketched together in the second diagram.

But at least one can argue as follows. When the water surface is tilted one can be sure that the water is being accelerated. There may perhaps be some upward or downward acceleration but there certainly is a horizontal component of acceleration and its direction is towards the place where the surface is lowest.

## A man on a raft

What would be the experiences of a man lying on a raft in a heavy sea; should he feel any tendency to roll off the raft as it tosses and tilts? If the man found any tendency to slide across the raft, or to roll across it, this would be equivalent to saying that the apparent

Fig. 1.9. The man on a raft.

direction of gravity, the 'false vertical', was not perpendicular to the surface of the raft. It seems likely, however, that the raft would always set itself parallel to the surface of the surrounding water and move to and fro in just the same way as that water. It should follow that the 'false vertical' would always be perpendicular to the raft, and the passenger would feel no tendency to slide or roll. He might be washed across the raft by broken water, but this is a different matter.

If you made a small flat raft with a rim around it, it would be interesting to see whether a ball on the raft would roll to and fro or whether it would stay in one corner all the time, when this raft was set floating in waves†.

## How to think about physics

Let's leave the breaking wave. It is too complicated; everyone finds it too complicated to be argued out in detail. It would seem much easier to argue about a wave pattern that did not change so rapidly as the breaking wave. A pattern that never changed would perhaps be the easiest to explain; let's look for it.

This is indeed how a physicist works, by trying to reason out the easiest problems first. Indeed he has to invent them. Think of it in this way. The world is a very confusing place and a physicist

† I have not tried this experiment, though I should like to know what happens.

tries to see patterns of order and reason in all this confusion. His technique is *never to argue about the world as it really is*.

Instead, he invents pictures that are much simpler than the real world and proceeds to reason about these. You know well enough that there never was a rigid body or a weightless string or a frictionless surface, but don't think of these as ideas invented merely to make physics easy for beginners. They are early triumphs of the physicist's art, by which he ' idealizes ' the world till it becomes simple enough to argue about.

He must remember of course that the real world won't behave in quite the same way as this idealized model suggests. His predictions will not quite agree with what actually happens. But they may be close if he's clever. So don't worry if it seems that you can't reason precisely about a real-world situation such as the way in which a knife cuts butter, or the way in which a bird flies or the way in which water waves move ; no one can do so, exactly.

But by reasoning about your ' idealized ' model you may see possibilities that had not occurred to other people. Then you make experiments to see whether they really happen. This is called ' research '.

*Ideal waves that never change*

On calm days on the coast one may sometimes be struck by the regular way in which long lines of wave crests follow one another in succession. Out at sea the water may seem quite glassy and flat, but nearer the coast each wave begins as a long faint shadow that becomes

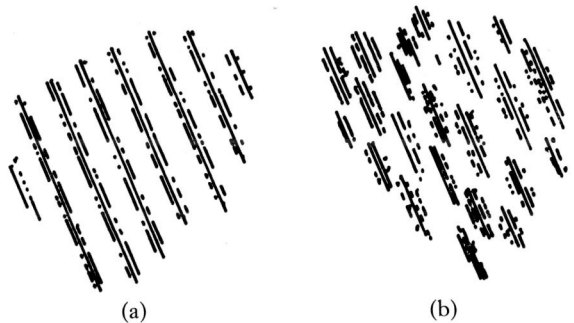

(a) (b)

Fig. 1.10. Ideal waves (*a*) and real waves (*b*).

more clearly a crest line as it approaches the shore and it is followed by another and another in apparently endless succession. Each grows in height as it advances, till one after another they break on the beach. This is a more promising picture. If one could imagine such waves in the open sea where there was no coast to interrupt them, these might be the simple unchanging kind of wave that we should consider first.

Waves in the open sea rarely have such a simple appearance. They are usually irregular in outline, each one different from the rest, and they have the disconcerting habit of disappearing as one tries to follow the crest, while new crests slowly grow that were not evident before. Nor do the waves usually have long parallel crests. When one looks down on the sea from an aircraft one can often see the waves but they usually have crest lines no longer than two or three times the 'wavelength' (the distance between successive crests). Still, regular trains of long-crested waves do sometimes occur at sea. When they are seen it is usual for them to have travelled a thousand miles or more from the storm that originally caused them. The sight of long parallel crests at sea can be very striking, but this is not the reason for our interest at the moment. We have picked on this kind of wave because it seems simple enough to argue about.

## An endless train of waves

We are thinking now of a pattern of long wave crests and long wave troughs that follow one another in endless succession. The waves are all alike and all equally spaced. They don't change their appearance as they travel. I will call this a 'regular train of long-crested waves'.

I am going to ask myself questions about how the water moves in such waves, both on the surface and underneath, and I also want to decide, if I can, the speed at which such waves would travel.

There will be very little mathematics in this chapter. I will keep the mathematics till later. Mathematics can be very useful to a physicist but one can often get a very good idea of what is likely to happen without putting everything down in symbols and equations. I myself think that it is very useful to do this. Mathematics I look on as a kind of machine, and a very nice one when you are quite sure how to drive it. But it can do only what you tell it to do. If you tell it wrong things it gives wrong answers; very precise answers of course but quite wrong all the same. I think that in physics it pays to argue first about what is likely to happen. When you have found the answer in this way, you set about 'proving' it by mathematics.

So let's see what ideas we can find.

The water goes up and down, that's obvious.

It also goes to and fro horizontally. My reason for saying this is that the water surface is tilted. We have already seen that when a water surface tilts, this shows that the water is being accelerated. It is not sufficient for the acceleration to be directly up or down; this would not cause the surface to tilt. There must be a horizontal component in the acceleration.

## The acceleration pattern

The tilt in front of a wave crest is in the direction that implies a forward acceleration, one in the direction in which the waves are going. Behind the wave the tilt is the opposite way and the acceleration (or the horizontal component of it) must be backwards. Do you agree?

So far as up and down motion is concerned, it seems obvious that water near the wave crests must be accelerating downward and water in the troughs must be accelerating upwards. So in fig. 1.11 I have sketched the pattern of acceleration.

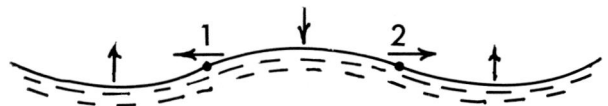

Fig. 1.11. The pattern of acceleration.

## *The velocity pattern*

We can now see the kind of motion that must be going on. We assume that all the surface particles go through the same motions but do so at different times as the pattern of motion advances. What is happening to particle 1 in the diagram will shortly happen to particle 2 when the waves have advanced by that distance. What is happening to particle 2 has already happened to particle 1 a little time before. So particle 1 has already been through the experience of a forward acceleration and now is accelerating in the backward direction. It was previously gaining forward velocity but is now losing it. Somewhere between these times, while it was above its average level, it had its greatest forward velocity. Apparently the water near the wave crests is moving forward. The same kind of argument shows that the water in the wave trough is moving backwards.

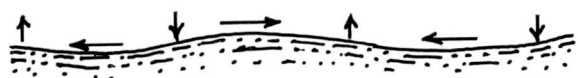

Fig. 1.12. The pattern of velocity in a wave travelling to the right.

Just to split a hair, notice that the particle might have, in addition, a steady forward drift or a steady drift backwards. Acceleration tells us only about the changes in velocity. But we need not worry about this drift. We take it that all the surface particles go through the same cycle of motion. If one particle has a steady drift, then they all share it. If there were a steady drift we could always take a new point of view in which we moved forward with the drift. Then we should see only a to and fro motion. This is the point of view that we are taking now.

We can therefore sketch the velocity pattern. It is forward on the crests, downward after they have passed, backward in the trough and upward as the next crest approaches. I have shown it in fig. 1.12.

*Motion below the surface*

It doesn't necessarily follow that the particles deeper down in the water do the same sort of thing as those on the surface. All sorts of complicated motions are possible in water. For example, in river estuaries it often happens that the less salty and lighter river water floats on the top of sea water and when the rising tide pushes the salt water inwards up the estuary the river water on top can be moving the other way, out to sea; the 'tide' is flowing opposite ways at the top and bottom. But in our wave problem I am assuming that the water is all salt or all fresh. There is no obvious reason for such differences in behaviour.

I shall therefore assume that the deeper water cycles in somewhat the same manner as the surface water. This suggests a pattern of velocity like that in the diagram of fig. 1.13.

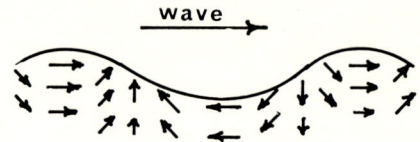

Fig. 1.13. A guess at the motion below the surface.

*Waves in shallow water*

It is not evident from our arguments that the to and fro motion should be just as large as the up and down motion. Indeed, if the water in which the waves are moving is quite shallow, one can see from fig. 1.14 that the horizontal motion will be the larger of the two.

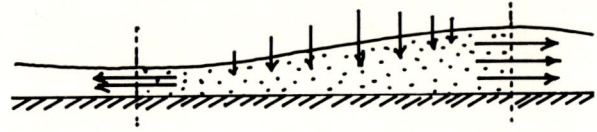

Fig. 1.14. Comparing vertical and horizontal velocities when the water is shallow.

Everywhere between the crest and the bottom of the following trough, surface water is in some degree moving downwards. The water must be escaping somewhere. Some is escaping forward under the crest, and some is escaping backwards under the trough. If the depth of water is much less than the distance between trough and crest it seems inevitable that the forward and backward speeds must be considerably greater than the downward speed. If the velocities to and fro are greater, the distance travelled to and fro will be greater. It seems that the paths of the particles will be like long ovals and they will be longer and narrower when the wavelength is greater and the depth is less.

We might estimate the value of this ratio. The distance between crest and trough is half the wavelength, or $\frac{1}{2}\lambda$ if $\lambda$ is the wavelength. The downward speed is greatest, however, about halfway between crest and trough and is weaker near the crest and trough. As a rough estimate of the downward flow we might multiply the maximum downward velocity by half the actual distance, $\lambda/4$. The boundaries over which the water is escaping (the depths under crest and trough) add up to $2D$ if $D$ is the average water depth. It seems that the maximum downward speed and the maximum horizontal speed are likely to be in the ratio $2D/\frac{1}{4}\lambda$ or $8D/\lambda$. The distances travelled vertically and horizontally by each water particle will also be in this ratio.

We have not argued very exactly. If the vertical component of velocity follows a sine law, the velocity ratio turns out to be $2\pi D/\lambda$. You can probably show this yourself.

*The speed of the waves*

We can now make a reasonable guess at the speed at which the waves should travel.

I will use $A$ to represent the distance that a particle on the surface rises above or falls below the mean level. This is called the 'wave amplitude'; the wave height from crest to trough is then $2A$.

I also need to use the distance that a particle travels to and fro horizontally from its average position. I will call this distance $B$. This amplitude is not necessarily the same as the amplitude $A$. We have just seen that if the wavelength is much greater than the depth of the water, then $B$ will be greater than $A$.

The formula we find for the wave speed will use these three lengths, $A$, $B$ and $\lambda$.

In making the following argument I shall suppose that the waves are 'low' in the sense that the distances moved by a water particle are quite small compared with the distance between successive crests. When a mechanical system makes quite small oscillations about a equilibrium state it usually happens that the motion is free from jerks or sudden changes; the motion usually approximates to 'simple harmonic motion'. I shall suppose that this is true of low water waves.

First, if we know the wavelength $\lambda$ and the vertical amplitude $A$, we can estimate the tilt of the water surface. If the wave outline were a zig-zag with sharp crests and troughs, the wave slopes would be tilted at an angle of $4A/\lambda$ (radians) to the horizontal†. But wave

† This would be the tangent of the angle, but the tangent and the angle itself are almost the same when the angle is small.

crests and troughs are usually rounded. This makes the tilt smaller near the crests and troughs and rather larger midway between them. Indeed if the outline of the low wave is a 'sine' curve, and I shall assume that it is, the maximum tilt turns out to be $2\pi A/\lambda$. Perhaps you can prove this.

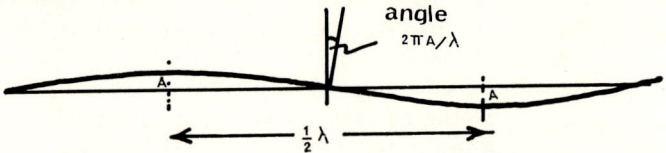

Fig. 1.15. Arguing the speed of travel of a wave.

The next point is that where the water surface is tilted it means that the false vertical is tilted, since the two are always at right angles. We can therefore find the horizontal acceleration of the water. You can see from the way in which one constructs the false vertical that

$$\frac{\text{horizontal acceleration}}{\text{gravity acceleration } (g)} = \text{tangent of angle of tilt}$$

$$= 2\pi A/\lambda.$$

So the horizontal acceleration, at its greatest, is $2\pi Ag/\lambda$.

In making this argument I have ignored any vertical acceleration of the water, thinking it likely that when the horizontal acceleration is greatest the vertical acceleration is likely to be near to zero.

We have found the greatest value of the horizontal acceleration. Of course the actual acceleration of a particle varies between $2\pi Ag/\lambda$ and $-2\pi Ag/\lambda$ as the particle waves to and fro.

But I assumed that we knew the distance $B$ that the particle moved to and fro horizontally. If one knows both the amplitude and the maximum acceleration in simple harmonic motion one can deduce the frequency. You may remember that if the 'angular frequency' is $\omega$, and if the amplitude is $B$, then the maximum acceleration is $\omega^2 B$. So

$$2\pi Ag/\lambda = \omega^2 B.$$

This allows us to write the value of $\omega$. If we want the periodic time $T$, rather than the angular frequency, it is $T = 2\pi/\omega$. If you do the algebra you will find that the period is given by:

$$T^2 = 2\pi \left(\frac{B}{A}\right)\left(\frac{\lambda}{g}\right) \quad \text{or} \quad \lambda = \frac{gT^2}{2\pi}\left(\frac{A}{B}\right).$$

Now we can find the wave speed, which I will call $c$. It must of course be given by:

$$c = \text{wavelength}/\text{wave period} = \lambda/T.$$

I can use the previous expression to substitute for $\lambda$ and when I do so the formula for the waves speed turns out to be:

$$c = \frac{A}{B} \cdot \frac{gT}{2\pi}.$$

Alternatively I can substitute $\lambda/c$ for $T$ and you will find that the formula becomes:

$$c = \left(\frac{A}{B} \cdot \frac{g\lambda}{2\pi}\right)^{1/2}.$$

These formulae show how the wave speed depends on $\lambda$ or $T$. They look rather complicated, but it would seem that waves with a long wavelength (large $\lambda$) or a long period (large $T$) will travel more quickly than shorter waves.

The formulae as they stand are rather unsatisfactory because they include the amplitudes $A$ and $B$ and we may not know what these are. One can have waves of different amplitude of course. It seems likely, however, that if the wave motion is made more intense, so that the particles move up and down through a greater distance, then the to and fro motion (the amplitude $B$) will increase too, and perhaps the ratio $(A/B)$ would remain the same. This argument suggests that so long as waves are not too steep (our arguments begin to be inexact if the tilts of the water surface are large), then the speed of travel of the waves does not depend on the wave height.

Later, in Chapter 3, we shall find out more about the value of the ratio $(A/B)$. I will close this chapter by looking at two special cases, that of very shallow water and that of very deep water, and go on to make some guesses about what the motion is like at places below the surface.

*Very shallow water; tides*

We call the water 'very shallow' when $D$, the water depth, is very much less than the wavelength $\lambda$. We have already guessed that in this situation the ratio $(A/B)$, that is the ratio of vertical amplitude to horizontal amplitude, would be small and about equal to $2\pi D/\lambda$.

If we substitute this value in the second equation for the wave speed the result is:

$$c = (gD)^{1/2}.$$

This is rather curious; the wavelength has disappeared from the formula and the wave speed depends only on the depth of water. All waves (if sufficiently long compared with the water depth) travel at the same speed, which depends merely on the water depth $D$.

Speed-boats moving in shallow harbours or estuaries are affected by this 'critical speed'. It is not difficult to make the boat move at a lesser speed, but when the critical speed is reached, the waves created by the boat build up into a large wave just ahead of it. The boat is driving uphill all the time and the going becomes more and more difficult. One hears of the difficulty in making a low-powered aircraft ' break the sound barrier ', that is travel faster than waves in air, and the speed-boat in shallow water has a similar difficulty in getting beyond the critical speed. It can be done best if the boat is

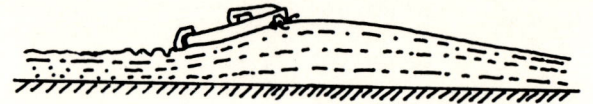

Fig. 1.16. Difficulties at the ' critical ' speed which is $\sqrt{(gD)}$.

capable of a sudden burst of power that carries it quickly beyond the critical speed before the big bow wave has had time to form. I will talk more about ship wakes in Chapter 7.

You can easily work out the critical speed. For example, suppose the depth of water is 6 m. The value of $g$ is 9·8 ms$^{-2}$. Then the critical speed is $6 \times 9·8$ or 7·8 ms$^{-1}$ or 29 kilometre per hour.

Tides are a kind of wave, though a very long one, so long indeed that even the deepest oceans can be called ' shallow water '. Let's find how fast tides travel.

In the open ocean the depth is about 4000 metres.

The value of $g$ is 9·8 ms$^{-2}$.

It follows that the speed of tides in the open ocean is about 200 ms$^{-1}$, about 700 kilometre per hour.

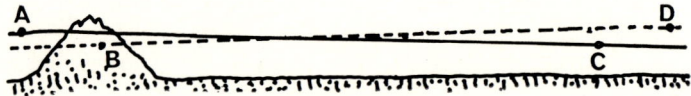

Fig. 1.17. As particle A moves to B and particle C moves to D, the crest of the tide wave advances from A to D.

This is not the speed of the *water* of course. It means that when the crest of the tide wave has reached one place, so that there is ' high water ' there, then after an hour there will be high water at a place 700 km farther on.

In the same interval of 1 hour the water may have moved perhaps 100 metre horizontally or 1 centimetre vertically. If we had been thinking of the ' low water ', the trough of the tide wave, the water would have moved the *opposite* way to the travelling tide wave. As I said earlier, a wave is a pattern of behaviour ; the pattern travels along, and very quickly in this case, but the water stays more or less where it is.

*Tsunamis*

Earthquakes in the earth below the sea sometimes cause a large area of the sea floor to rise or fall. The change can be very sudden. On many occasions, the log-book of a ship reports that while it was over deep water the ship struck a rock. This was the only interpretation that the captain could give of the sudden severe shock to the ship. Yet later navigators found that no rock was there. Presumably the shock came from an undersea earthquake.

Sometimes an earthquake shifts the sea floor permanently. If the sea floor rises, the sea rises too, but then of course it begins to pour away into the surrounding water. Or if the sea floor falls then water surges inwards from surrounding regions. This motion of the water is so widespread that a ship in the region would not notice it; it is a matter of the whole sea moving. The effects spread to a great distance, somewhat as ripples spread outward from a stone thrown into water, but the water waves that spread from an earthquake would not be noticed by any ocean traveller on deep water, for they are so very 'low'. The rise and fall of the water in such waves may be only a foot or two but the undulations have a wavelength of so many miles that the disturbance is quite unnoticeable in the open sea.

These very long water waves travel almost at the speed of the tides. The deep oceans are shallow compared with their wavelength. The menace of such events is seen when waves reach some distant shore, as they speedily do. You may have noticed that ordinary sea waves become higher as they move in from the sea to a shallow beach. They grow higher and finally break. A similar increase in height occurs with these water waves from earthquakes but the increase can be more marked. The long wave can arrive at the land as a disastrous wall of water 10 or more feet high that can sweep over low-lying country.

Fortunately, these very destructive water waves from earthquakes are rare in most well-populated countries. The islands of Hawaii are troubled by them rather frequently and Hawaii is the centre of a communication system that has been set up to give warning of such waves to countries that border the Pacific Ocean.

The water wave from an earthquake used to be called a 'tidal wave' but you can see that this is not a well-chosen name. Its cause is not the same as the cause of the tides. Today such a wave is usually called a 'tsunami'. It is a Japanese word, and both earthquakes and the water waves from them are not unusual in that country.

Tsunamis usually take the form of several waves in succession at intervals of 10 or 20 minutes, all coming from the same original disturbance. The leading wave is not always the greatest. It frequently happens that the first sign at a coast is a depression of the sea level. If you should ever notice that the sea has receded unusually quickly, my suggestion is that you do not follow it at once to explore the new rock-pools, but go to high ground in case a tsunami should follow in the next 10 minutes.

*Deep-water waves*

We are thinking here of a train of waves moving over water whose depth is much greater than the separation between crests (the wavelength).

When the waves were in very shallow water we argued that the ratio $A/B$, the ratio of vertical and horizontal motion in the water, depended on the ratio of wavelength to depth.

In deep water we were not able to guess the ratio $A/B$. Yet if the water is very deep we can think of wave-trains with different wavelengths $\lambda$ and the ratio $\lambda/D$ would be zero for all of them. It seems reasonable to guess that in deep water the ratio $A/B$ will be the same whatever wavelength we choose.

Then our formulae for the speed $c$ say that wave-trains with a longer wavelength or a longer period travel at a greater speed.

This actually happens. A storm in the middle of the open ocean causes a commotion in which the wind produces waves of all sorts of lengths. Those with a long wavelength (which also implies a long period) travel more quickly away from the storm and they are the first to reach the shores of the ocean. Waves of shorter period travel more slowly and arrive later. Indeed if one keeps watching the waves that arrive at a coast one can see this happening. The first sign that there has been a distant storm at sea, perhaps 4000 kilometres away,

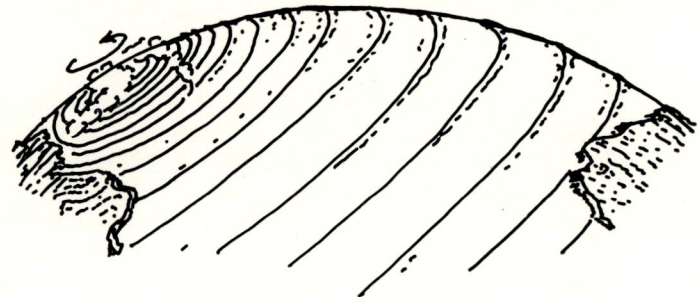

Fig. 1.18. Waves spreading out from a storm at sea. I could not draw them all. Think of 1000 times as many as are shown.

is the arrival of breakers with a time interval of perhaps as much as 20 second between each breaker and the next. Some hours or days later, depending on the distance of the storm, the intervals may have become 15 second or 10 second; the shorter period waves have arrived.

It is possible in this way to calculate the distance of the storm that caused the waves, but we will look at this in Chapter 5.

We shall find out later that for waves in deep water the ratio $A/B$ is actually 1. Our formulae for wave speed then become:

$$c = gT/2\pi = (g\lambda/2\pi)^{1/2},$$

and these are indeed correct. We will leave this however till we have justified it in Chapter 3.

## Motion below the surface

We haven't said much about this. We shall come to definite conclusions about it in Chapter 3. I said, however, that in physical problems it was alway a good idea to use simple non-mathematical arguments first of all.

One can see very easily that the up-and-down motion in waves must be *less* when one looks at places below the surface.

Think of it in this way. When water is carried up and forward in a wave crest it must become shorter in the horizontal direction. Since it doesn't change its volume it must have become longer in the vertical direction. This must mean that the distance risen by the top of the piece of water is greater than the distance risen by the water below. The rise and fall must be less in the deeper water.

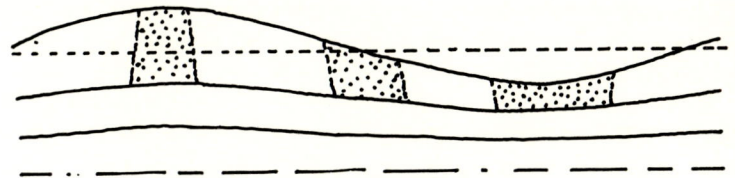

Fig. 1.19. The motion must be less lower down.

The argument depends on the water becoming narrower horizontally when it is in a wave crest. I think that you will agree that this happens. Water in front of a wave crest has been moving backward in the trough that has just passed. Water behind a wave crest has been carried forward while the crest was passing it. These two pieces of water are now closer together than they would have been if there had been no wave. So the water in the wave between them has been compressed in the horizontal direction.

So wave motion becomes weaker at greater depths. The deeper you go, the less wave motion you feel.

## Conclusion

You now see that you can reason about waves in quite a simple way. But two more ideas are needed to make the story complete. These are 'circulation' and 'velocity potential'. I talk about them in the next two chapters, and then go on in the rest of the book to look at the curious patterns one can make by adding straight wave-trains together.

# CHAPTER 2

## the motion of a frictionless fluid

Do you like mathematical ideas?

This chapter and the next show some rather curious mathematical ideas about fluid motion. I include them because I want to show you the way in which physicists argue about fluid motion. I will do my best to make the arguments sound like common sense, but if mathematics disagrees with you I suggest that you skip these two chapters and read on from Chapter 4.

*Inventing the idea of a fluid*

I have a jar of water sitting quietly on the table in front of me. The books tell me that water is a collection of molecules colliding and rebounding at speeds like a kilometre a second.

The idea may be true. I am willing to believe it. Yet, when I think about waves the idea of molecules doesn't seem a useful one.

Fig. 2.1. It stays where it is.

I drop a small crystal of dye into the water. It sinks to the bottom and soon dissolves to form a small patch of stained water. Molecules or not, I propose to say that the patch of stained water 'is at rest', for I can see that it stays where it is. It certainly doesn't dart about the jar at the speed of a kilometre a second.

In thinking about the motion of water I shall argue as if it were possible for me to introduce, at any place, a drop of stained water and watch it move. I shall say that this shows me the motion of the water. I shall forget all about molecules.

In thinking in this way I have to ignore another, quite real, effect. Any small patch of dye in my jar would slowly expand until the dye had extended throughout all the water. And water does what the dye does; water continually 'diffuses' through water. But this process is very slow. It would take several days to see it happen in my jar. In thinking about waves I shall forget this process.

There is still another effect that I shall ignore. I shall suppose that water is frictionless. This is certainly not true. You can stir water in a jar and leave it rotating like a whirlpool when you take the

Fig. 2.2. Impossible if water had no viscosity (fluid friction).

spoon out. Without fluid friction this would be impossible. But the kind of motion that occurs in water waves leads to very little friction. I shall ignore friction entirely.

The absence of friction means that where two pieces of fluid are in contact and pressing on one another, the force that each exerts is at right angles to the surface of contact. You will have met this idea in hydrostatics. I shall take it to be true even when the water is moving.

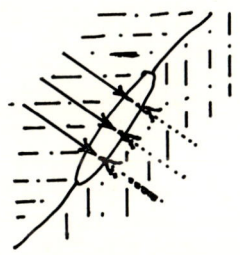

Fig. 2.3. The idea of pressure, thrust per unit area.

You will also have met the idea of 'pressure'. Each small portion of the contact surface transmits its own quota of force, an amount proportional to its area. The constant of proportionality, the 'force per unit area', we call the 'pressure'†.

† In a frictionless fluid there is only one number for the pressure at any particular place and time. Pressure doesn't depend on direction. But if there is fluid friction, every place has three different pressures and three different frictions at the same moment, so fluid friction is a good topic to avoid.

21

In case you should feel disappointed at my ignoring the molecules and the friction, remember that no physicist can ever argue about the world as it really is. He always needs to 'idealize' it. Half the fun in real-life physics is in ingeniously falsifying the real world until the picture becomes simple enough to argue about.

## 'Circulation' in a frictionless fluid never changes

I want to prove this, but I had better first illustrate what I mean.

Think of water rotating symmetrically in a circular basin, which is fitted with a central waste-pipe so that water continually escapes. If you think in particular of the particles of water that happen to lie on

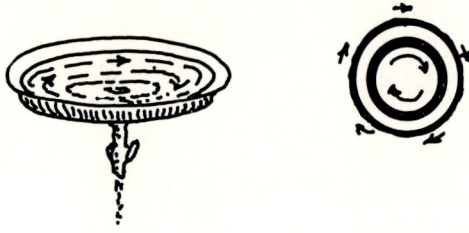

Fig. 2.4. It speeds up as it moves inwards.

some circle, symmetrical in the basin, you can see that as well as moving round the circle they will all creep towards the centre, and the circle they make will gradually shrink in diameter.

In this ring of particles the 'circulation' means the tangential velocity (which by symmetry is the same all round the circle) multiplied by the distance round the circle.

The theorem I wish to prove says that if friction can be ignored, then the numerical value of the 'circulation' in this ring remains always the same. By the time that the particles of water have closed in so far as to make a circle of half the original diameter, then the *tangential* velocity must have become twice what it was at first. The distance round the circular path being less, the tangential speed must be greater in proportion, in order to keep the 'circulation' the same.

I have not yet proved the theorem of course, but in this particular example you could justify it by arguments that you already know. If you think of the particles as forming a narrow annulus or ring, whose inner and outer surfaces are circular, then the forces on this ring due to the water both inside it and outside, will all be directed at right angles to these boundary surfaces, in fact along the radii. The forces on this ring of water have no torque about a vertical axis through the centre. So the angular momentum of this ring of water will not change. As its diameter gets less its angular velocity must increase. If you work this out you will be able to prove that when the diameter has become half of what it was, the tangential speed will have doubled, as I said.

But it turns out that 'circulation' remains the same *whatever the shape of the loop you take and for any kind of water flow*. I want to prove it in general.

## The circulation theorem

Imagine that in a moving fluid you have picked on a long filament of particles that form a closed loop. It need not be a circle; it need not even be a flat loop.

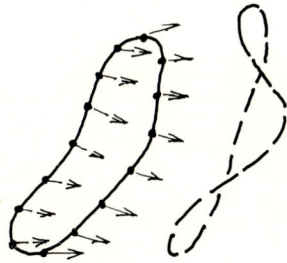

Fig. 2.5. A loop of water particles drifting with the rest of the water.

This loop of particles will drift along with the moving fluid, probably stretching and twisting as it moves. What do we mean by the 'circulation'? We have to consider everywhere the component of velocity that is directed along the loop. It is clear that this component may well have different values at different places round the loop. Indeed we shall have to pick one sense in which we might move round the loop and call it the positive sense. If at some place the tangential component of velocity happens to be directed the other way, then we shall call it negative.

Because the tangential velocity isn't the same everywhere we must consider separately all the small pieces of the loop. Imagine if you like that you have chosen a large number of particles closely spaced all round the loop and have marked them in some way so that you can identify them. We can now consider all the various small elements of the loop, between each marked particle and the next.

## The 'flow' along an element

Consider one small element of the loop of fluid. Because it is so small we can think of it as straight. We can also assume that all the particles in it are moving in almost exactly the same way.

If the length of this element is $\delta l$ and the component of fluid velocity parallel to it is $u$, then the product

$$u\delta l$$

is called the 'flow' along this element.

If we calculate the flow along each element of the loop, then the sum of all these flows is called the ' circulation ' round the loop.

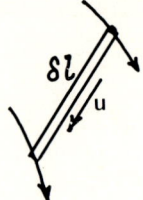

Fig. 2.6. ' Flow ' along an element of the loop.

I am going to show that the flow along different elements can change with time as the fluid moves, but that the total of the flows, the circulation, always remains the same.

*How the flow can change*

The particles of fluid may be accelerating, changing their velocity. Then, of course, the value of the flow will change. I will look into this later. But even when there are no accelerations, the flow in an element can change, and I will look at this effect first.

*Velocity*

Suppose first that the particles happen to be moving parallel to the length of the element. They will not all have exactly the same

Fig. 2.7. Lengthwise velocity can cause a change of ' flow '.

velocity. The particle at one end of the element may have a velocity $u$, while the particle at the other end has some velocity $u + \delta u$, the intermediate particles having intermediate speeds†.

† A mathematician would say that I am assuming that the velocity is a ' differentiable function of position '. Things need not be like that, I suppose. Perhaps the water might be full of strong eddies so microscopically small that no matter how small an element I took, I would find that the different particles moved in very different ways. I shall not try to argue what would happen then. My business as a physicist is to invent some situation that I *can* argue about, not one that I *can't*.

Now you can see that the flow is changing, not because the velocity is changing but because the length of the element is changing; its two ends are moving farther apart.

The length would increase by an amount $\delta u$ per second.

To get the increase in flow, multiply by the velocity. It is good enough to use the average value, which is $u + \frac{1}{2}\delta u$.

So the increase in flow per second is:

$$(u + \tfrac{1}{2}\delta u)\,\delta u.$$

If you look hard at this you will see that it is just half the difference between the *squares* of the velocities at the two ends:

$$\tfrac{1}{2}[(u+\delta u)^2 - u^2].$$

I have been using the symbol $\delta$ to mean the increase in a quantity as one moves from one end of the element to the other, going in the positive sense. Then writing $u$ for the tangential velocity, which will vary as one moves along the line, I can write:

$$\text{rate of increase in flow} = \tfrac{1}{2}\delta(u^2).$$

But now consider a different case in which the particles all happen to be moving in a direction at right angles to the length of the element. Our problem is in three dimensions, so I need to say, in addition, that I am thinking of the particles as all moving in the *same* direction.

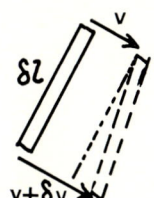

Fig. 2.8. Transverse velocity can cause a change of 'flow'.

In this case the element has zero flow at first. But a flow soon develops; the reason is that the element changes its attitude as the particles move. At one end the velocity may be $v$ and at the other end $v + \delta v$. Then in unit time one end will move out more than the other by a distance $\delta v \times 1$ second. The direction of the element will have changed by a small angle that I can write as $(\delta v \times 1\ \text{second}/\delta l)$ in radian measure.

To find the component of velocity along the element in this new position, multiply this angle by the average velocity $v + \frac{1}{2}\delta v$.

Then to get the flow multiply by the length $\delta l$. Notice that the slight increase in $\delta l$ is a second-order effect and can be ignored. The result is that, per second, the flow increases by:

$$(v + \tfrac{1}{2}\delta v)\,\delta v.$$

Once again this can be written as half the difference between the square of the velocity at the two ends:

$$\text{rate of increase of flow} = \tfrac{1}{2}\delta(v^2).$$

In general, of course, the velocity of the particles will not be solely along the line nor solely at right angles. It will have components, say $u$ along the line and $v$ and $w$ in directions at right angles. But I hope you will agree that in this case we can just add together the effects that follow from each of the components taken separately. Then the whole rate of change of flow is:

$$\tfrac{1}{2}\delta(u^2) + \tfrac{1}{2}\delta(v^2) + \tfrac{1}{2}\delta(w^2).$$

By Pythagoras' theorem the sum of the squares of the rectangular components is just the square of the total speed of the particle, which I will call $q$:

$$\text{rate of increase of flow in an element} = \tfrac{1}{2}\delta(q^2).$$

This is the effect of velocity.

Before going further I will point out how it can come about that the flow may be changing in all the elements of the loop, yet the total flow, the circulation, may remain the same.

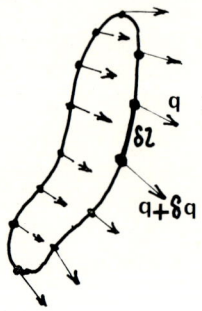

Fig. 2.9. Yet the sum of all the flows (the 'circulation') can stay the same.

If we move around the loop looking in succession at its various elements, we may at first be moving into regions where the fluid speed $q$ is greater. Then for these elements $\delta(q^2)$ will be positive, showing that the flow in them is increasing. But we have chosen to consider a complete loop so that we finally return to the original point where $q$ has its original value. Necessarily, therefore, if $q$ first increases as we move round the loop there must come a stage at which it begins to decrease. Then $\delta(q^2)$ turns negative, showing that the flow in those elements is decreasing. The sum of all the rates of change of flow is just the sum of all the quantities $\tfrac{1}{2}\delta(q^2)$ and since we arrive finally at the same value of $q$ the sum of all these increments in $q^2$

must of course be precisely zero. The *circulation* has zero rate of change, it remains always the same.

*Acceleration*

Now that we have seen how velocities affect the flow I will look at accelerations and ignore the effect of velocity. To do this I will imagine that the velocities of the particles are all zero at the *middle* of the short time interval I am considering. On the other hand, I will think of the accelerations as being constant during this interval.

In such a situation you will agree, I think, that every particle reaches, at the end of the interval, exactly the same position as the one it had at the start. For instance, if the acceleration of a particle is say $a$ in some direction, and if its velocity is zero in the middle of the interval then the velocities at the start and the finish will be just equal and opposite, say $-\frac{1}{2}a\delta t$ and $\frac{1}{2}a\delta t$ if $\delta t$ is the length of the interval. The average velocity during the interval is zero and consequently the overall displacement is zero.

The length of the element is just the same at the end of the interval as it was at the start, and so is its orientation. Changes in length and orientation were the key ideas when we considered velocities, but when we consider accelerations it is not changes in length or orientation that lead to changes in flow. The only factor that need be considered is the change in velocity, and there is one, of course. If the acceleration happens to be along the length of the element, and equal to $a$, then this is the change in velocity per unit time, and the change in flow per unit time is :

$$\text{rate of change of flow} = a\delta l.$$

Accelerations perpendicular to the length of the element do not affect the flow, of course, because they do not alter the component of velocity parallel to the length.

What is the acceleration of the fluid in the direction parallel to its length ?

If the fluid accelerates, it does so because of forces acting on it. Forces from the surrounding fluid are one reason for acceleration. I will consider them first.

Think of an element of the loop as being a thin cylinder of fluid of length $\delta l$ and cross-section $A$. If the fluid is frictionless the forces on the sides of the cylinder act at right angles to its length. They may produce accelerations in a direction at right angles to the length but these do not directly affect the flow.

Forces parallel to the length come from pressure on the ends of the cylinder. The pressure may be $p$ at one end and $p+\delta p$ at the other. The section at each end is $A$, so you see that the overall force is negative (backwards) and equal to :

$$-A\delta p.$$

To find the acceleration we divide by the mass of this cylinder of fluid (density times volume), which is:

$$\rho A \delta l.$$

The increase of tangential velocity per unit time is therefore:

$$-\delta p/\rho \delta l.$$

Multiplying by the length $\delta l$ we find that

rate of increase of flow (due to fluid pressure) $= -\delta p/\rho$.

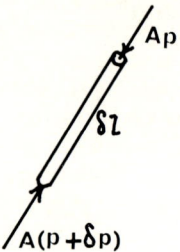

Fig. 2.10. Acceleration due to fluid pressure can change the flow in an element.

The density comes into the argument this time. In discussing water waves I am prepared to suppose that the density of the water is the same everywhere. The factor $1/\rho$ is just a constant number. Then once again we can see that the circulation round a complete loop of fluid will not change. There are likely to be different pressures

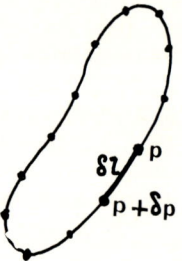

Fig. 2.11. Yet the total circulation stays the same.

at the two ends of all the various elements, but in moving round a closed loop we return to the same place where the pressure has its original value. So the sum of all the increments $\delta p$ is just zero; the total rate of change of flow is zero. Fluid pressure (in the absence of friction) cannot change the circulation round a fluid loop.

There are other forces on the water. I should consider gravity. Gravity acting by itself would produce a downward acceleration $g$ in all the particles. It is, however, the component of acceleration along the length of each element that matters. I had best write this as $g'$. It depends on the inclination of each element to the vertical, so $g'$ will be different in different elements of the loop.

If $g'$ is the acceleration along the length of each element, I think that you will agree that I can write:

rate of increase of flow in an element (due to gravity) = $g'\delta l$.

How shall I argue about this? We usually think of $g$ as an acceleration but it is equally true to think of it as the force that gravity exerts on an object per unit mass of the object. Similarly $g'$ represents the component of this force along the length of each element.

But force times distance moved is 'work'. The quantity $g'\delta l$ represents the work that gravity would do on some unit mass in moving it from one end of an element to the other.

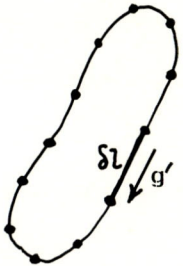

Fig. 2.12. Gravity acceleration changes flow in each element, but cannot alter the total circulation.

But in taking this unit mass all the way round the loop the work done by gravity would be just zero. Gravity is what we call a 'conservative' field of force. *The sum of all the quantities $g'\delta l$ is just zero.* The implication is that in the original loop of fluid the sum of all the rates of change is zero. Once again, the circulation round the whole loop remains unchanged†.

*Conclusion.* The argument has been rather long, but it seems that if, in a frictionless fluid, you fix attention on any closed loop of

† Imagine if you like that you have replaced the loop by a loop of frictionless wire held in that position somehow, and that all the other fluid has been taken away. Now you let a small object of unit mass slide on the wire, going in the positive sense round the loop. In some places you would see the object pick up speed, showing that gravity was doing work on it and that the quantity $g'\delta l$ was positive. In other places it would lose speed, showing that the work done was negative. The implication is, of course, that in the original loop of fluid gravity would tend to increase the flow in some elements and reduce it in others.

particles, then the loop may greatly change its shape and size as the fluid moves, but the 'circulation' round the loop will always be the same†.

Circulation is a curious idea and I wonder how Daniel Bernoulli (about 1738) ever came to think of it. But we shall put it to good use now.

*Irrotational motion*

Waves are a kind of motion that can spread into water that was previously at rest.

Water at rest has no circulation anywhere.

Fig. 2.13. Water pressure and gravity forces make waves advance into water originally at rest. They cannot produce circulation. Waves must be an 'irrotational' motion.

Waves spread by the action of water pressure and gravity. We have seen that these cannot change circulation.

So when waves spread into water that was previously at rest they must be a kind of fluid motion that has *no circulation anywhere*. It is an 'irrotational' motion. Let's look at this idea more closely.

In irrotational motion there is no circulation. Can we look at this in another way?

Take any two points in the water, say A and B. You can join these up by any number of different paths. But it seems that the total 'flow' from A to B is the same along every one of these paths if the flow is 'irrotational'.

For, you see, any two paths taken together form a closed loop. The total flow, the circulation, round the loop is to be zero. So the flow along the two branches must be equal and opposite. But we are then following one of the paths in the reverse sense, from B to A. If we always consider the direction from A to B as being the positive one we find that the flow along the two paths is exactly the same.

This reminds one of the mechanics of a gravitational field. If you have two positions, C and D, in a gravitational field and you want to move an object slowly from one position to the other, you can choose

† You can change the circulation by producing another kind of force. For instance if you pass an electric current through the water and have a magnet nearby, then the water will begin to move and circulation will develop. I suggest that you try this.

Or again, if you warm one side of a jar of water, its density changes and you get circulation, convection currents. But I shall take no notice of these possibilities when thinking of waves.

30

any path you like and the work you need to do is just the same. This gives rise to the idea of gravitational potential energy that depends only on the position of the object.

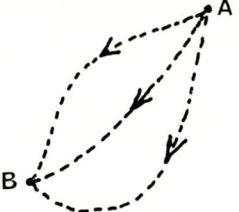

Fig. 2.14. The idea of a velocity potential when the water motion is irrotational.

*Velocity potential*

Similarly, in a frictionless fluid, we can invent an idea called velocity potential. This is a number that depends on position. The difference in velocity potential at any two points is just the flow along a path, any path, that joins the two points.

In gravity you can imagine two points C and D quite close together and join them by a straight line of length, say, $\delta l$. Then the work you need to do in pushing the object from C to D is just the excess of potential energy at position D. I will call it $\delta(\text{P.E.})$. If you need to exert a force $F$ parallel to the line then the work $F\delta l$ is equal to the increase $\delta(\text{P.E.})$. So you see that there is a relation between the force you must exert and the increment of potential energy. If you prefer to think of the force due to gravity, which is of course just the reverse of the force you must exert, you can say:

component of force due to gravity $= -\delta(\text{P.E.})/\delta l$.

In the fluid flow problem we use velocity potential in a similar way. People commonly use the Greek letter $\phi$ (phi) to represent it. It is a number that depends on position, and it may vary with time too. Whereas potential energy relates to force times distance, velocity potential relates to velocity times distance, so we have the rule:

component of fluid velocity $= -\delta\phi/\delta l$.

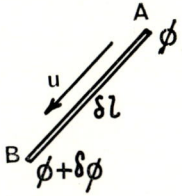

Fig. 2.15. How to deduce velocity from velocity potential. By convention we understand that the velocity is from high potential to low, so $\delta\phi$ in this diagram would be negative.

We can choose the elementary length δ*l* to be in any direction we like and we get the component of velocity in that direction. You notice that if we move along the length δ*l* and find that $\phi$ is increasing, so that δ$\phi$ is positive, then the negative sign means that the fluid velocity is backwards. The fluid is moving from regions of high potential towards regions of lower potential. This is the usual convention.

*Making up patterns of irrotational motion*

So it seems delightfully easy to make up patterns of fluid flow in which the motion is irrotational. All one has to do is to invent some kind of continuous distribution of numerical values in space and call this a velocity potential. Then the gradient, the rate of change with position, tells the speed of the fluid. *The result is certain to be an irrotational motion.* With luck it might even represent a wave. Let's try this.

I will use polar co-ordinates and use *r* to stand for the radial distance from some central point. I need not consider angles for I am going to try the effect of writing:

$$\phi = r^2.$$

The number $\phi$ is now defined for every place. It is just equal to the square of the distance *r*. Obviously $\phi$ grows larger as one moves outward. The fluid flow must be inwards (from high potential to low) and by symmetry you can see that the flow must be directly towards the centre. If you work out the fluid velocity along the radius it is:

radial velocity (in the direction of increasing *r*) = $-d\phi/dr$

$$= -2r.$$

The negative sign shows that the flow is inwards.

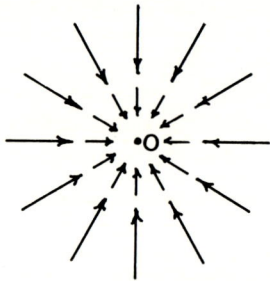

Fig. 2.16. One kind of irrotational flow; but it doesn't describe a wave on water.

The flow is inwards from every direction, so where is the fluid going? It is not escaping at the origin because the inward velocity, which is $2r$, falls to zero, where $r$ is zero. I think you may agree that this picture represents a uniform collapse, a sort of expanding universe in reverse. Every part of the fluid is becoming more and more compressed in a quite symmetrical way. Such a motion might occur in air because air is compressible. It won't do for water waves; I want to think of the water as being incompressible.

## Finding the rule that means no change in volume

We need to find some rule for choosing a velocity potential that does not require the fluid to alter its volume.

To find this I need to become a little more mathematical. You are of course acquainted with Cartesian co-ordinates, where one locates a point in space by choosing some origin and measuring component displacements in three directions at right angles†. I will denote the distances by $x$, $y$ and $z$. I will write $u$, $v$ and $w$ for the components of fluid velocity in these directions.

Imagine that you have marked out in space a small cuboid and that the fluid is moving through it or past it. The faces are perpendicular to the three co-ordinate directions and we can suppose that the lengths

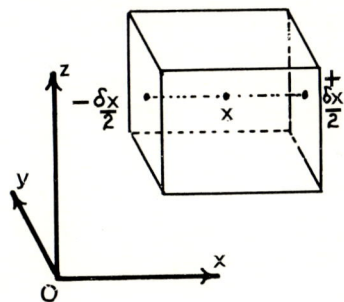

Fig. 2.17. Finding the rule that ensures no contraction or expansion.

of the edges are $\delta x$, $\delta y$ and $\delta z$. The very centre of the cuboid can be the position $(x, y, z)$ where the components of fluid velocity are $u$, $v$ and $w$.

We need to make sure that just the same volume of fluid flows into this cuboid as comes out of it. Otherwise some contraction or expansion would be taking place.

Think first of the two faces that are perpendicular to the $x$ direction. The amount of fluid that crosses these faces is determined by the

---

† This ingenious idea was invented by a Frenchman, Descartes, as a result of a dream on the night of 10 November 1619. He was a soldier aged 23 on active service at the time.

component of velocity in the $x$ direction. The other components of velocity are parallel to these faces and don't count. Consider one of the faces, the one nearer the origin. The fluid velocity won't be uniform all over it, yet I think that if it is a little larger at one corner of the face it will be a little less at the opposite corner. I think it is fair as an average to take the velocity existing at the centre of this face.

Now, this velocity will not be $u$, because I am supposing that $u$ represents the velocity (in the $x$ direction) at the centre of the whole cuboid, the point $(x, y, z)$. At the centre of this face the velocity may be a little less, say

$$u - \tfrac{1}{2}\delta u.$$

You will see in a moment why I have put the $\tfrac{1}{2}$ in.

When I look at the middle point of the other face, I think I may expect to find that the velocity there is greater than $u$ by just the same amount, namely

$$u + \tfrac{1}{2}\delta u.$$

This is because the centres of the two faces are just the same distance from the cuboid centre but on opposite sides of it.

Then the velocities at these two faces differ just by $\delta u$. This is why I introduced the factors of $\tfrac{1}{2}$, so that $\delta u$ means the increase in $u$ velocity that one sees on looking at a point displaced a distance $\delta x$ in the $x$ direction.

But the area of both these faces is $\delta y\, \delta z$. It follows that, through these two faces, fluid is escaping from the cuboid at a rate:

$$\delta u\, \delta y\, \delta z.$$

I go on now to look at the other faces. When I think of the pair of faces at right angles to the $y$ direction, only the component of velocity in the $y$ direction is important and I would, by a similar argument, find that fluid was escaping from these at the rate $\delta x\, \delta v\, \delta z$. The third pair of faces would give an additional rate $\delta x\, \delta y\, \delta w$. The total rate at which fluid is leaving the cuboid is just

$$\delta u\, \delta y\, \delta z + \delta x\, \delta v\, \delta z + \delta x\, \delta y\, \delta w.$$

I want to make sure that there is no expansion or contraction, but if I merely equate the last expression to zero this does not make a sensible mathematical statement. The quantity I have written is already very small merely because the cuboid itself is very small. I ought first to divide by the volume of the cuboid to show the *fraction* that escapes per second. Then I will say that this fraction is to be zero. So I divide by $\delta x\, \delta y\, \delta z$ and get the statement:

$$\frac{\delta u}{\delta x} + \frac{\delta v}{\delta y} + \frac{\delta w}{\delta z} = 0.$$

And finally, since these all refer to very small changes in position and velocity, I think you will agree that these ratios might be written as differential coefficients† :

$$du/dx + dv/dy + dw/dz = 0.$$

So this is the rule that the velocities $u$, $v$ and $w$ must everywhere obey if there is to be no expansion or contraction.

But we wanted a rule that would refer to the velocity potential and help us to choose it properly. It is easy to get this now. When we have a potential that depends on $x$, $y$ and $z$, we ask how it changes with $x$, not letting $y$ or $z$ change, and this gives the velocity $u$. I will write :

$$u = -d\phi/dx.$$

Then in our rule, the first term can be written :

$$-\frac{d}{dx}\left(\frac{d\phi}{dx}\right),$$

and we abbreviate this, as usual, to $-d^2\phi/dx^2$. The other two terms can be changed in a similar way. So after cancelling the negative sign throughout we get the rule as :

$$d^2\phi/dx^2 + d^2\phi/dy^2 + d^2\phi/dz^2 = 0.$$

In thinking about water waves we must make sure that our potential obeys this rule.

---

† Careful now! The velocity $u$ will usually depend on all three co-ordinates. If you had a formula :

$$u = 2x^2y + xz^2,$$

what would you understand by $du/dx$ ? One interpretation would be to treat the $y$ and $z$ as constant numbers and differentiate for $x$ only, getting :

$$du/dx = 4xy + z^2.$$

But is this what we intend ? If you look back at the argument you will find that it is. The change $\delta u$ arose through a change $\delta x$ without any change in $y$ or $z$. So the ratio $\delta u/\delta x$ does mean this kind of differential coefficient.

A mathematician calls it a 'partial' differential coefficient and to show this he usually writes it with a kind of curly d, $\partial u/\partial x$.

# CHAPTER 3

## the speed of a wave-train

I try to imagine a perfect wave-train. The waves are long parallel crests, all alike, and all advance at the same speed so that they pass me in regular succession. This is my ideal wave-train.

It never happens of course., Every wave at sea is different from every other. This perfect wave-train is my invention. But the art of making progress in physics is to invent such idealized pictures. I will hold on to my idea of a wave-train. You will see later how useful it is.

What questions shall I ask myself? They are:
*First*: how does the water move inside the wave-train?
*Second*: how fast does the wave-train travel?

The last chapter gave me a key idea; the water motion is irrotational, so I must think of a velocity potential, $\phi(xyz)$.

Then again, if the water is to be thought of as being incompressible, the last chapter showed me that this velocity potential must everywhere fit the rule:

$$d^2\phi/dx^2 + d^2\phi/dy^2 + d^2\phi/dz^2 = 0.$$

I shall make use of these two ideas.

*The idea of a ' steady state '*

Here is a useful trick.

I can imagine myself moving forward at a steady speed that just keeps pace with the waves.

Fig. 3.1. Taking the ' steady-state ' point of view.

Things look even simpler than before, if I do this. The wave crests and wave troughs no longer pass me; they seem to stay always in the same places, while the whole sea streams backward past me.

By taking this point of view I have created what people call a ' steady state '. You may actually have seen some situation like this, when a fixed rock makes waves on the surface of a fast-flowing river. The waves it makes stay in the same place while the water moves through them.

My ' steady-state ' picture makes the arguments simpler because the pattern of motion never changes. The velocity of the water is different at different places, but if I look at any one place (a fixed

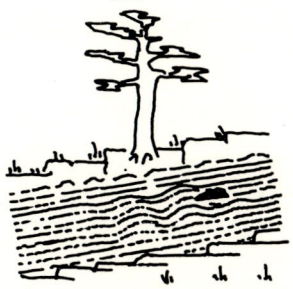

Fig. 3.2. Stationary waves on a river.

position relative to myself), the velocity at that place never changes. When I begin to think of suitable velocity potential, it will *not need to vary with time*. This makes things much simpler.

In most of this chapter I shall be thinking of this ' steady-state ' situation. I shall try to decide just how the water must move, and at what speed the stream must travel in order to hold the waves stationary.

But when I have solved this ' steady-state ' problem I think you will agree that it will be easy to go back to the original point of view, and to decide what things will look like to a person who sees the waves moving past him while the water, on the whole, remains in the same place.

*Arguing about the ' steady state '*

I will take it that the $x$ distances are measured forward horizontally in the direction perpendicular to the wave crests. The $z$ distances I will take as being measured vertically upwards. It follows that the $y$ distances are taken horizontally in a direction parallel to the wave crests.

How shall we describe the steady backward flow of water ? This is easy. I write :
$$\phi = cx.$$
Then the velocity $u$ in the positive $x$ direction is given by :
$$u = -\mathrm{d}\phi/\mathrm{d}x.$$
Obviously this is just
$$u = -c,$$
and all the water is moving backwards at the speed $c$.

But this formula represents only a stream with no waves. How can I include them? Notice first that if you look in the $y$ direction parallel to the crests, the motion remains the same however far you go.

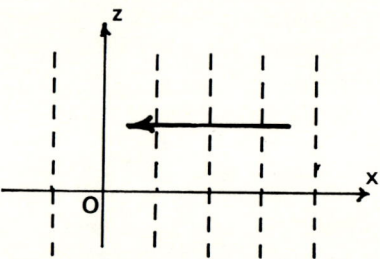

Fig. 3.3. The contours (broken) of velocity potential for a steady stream.

It seems unlikely that particles move in the $y$ direction so we need not include $y$ in the formula.

But we must include $x$. If we move in the $x$ direction we pass crests and troughs in succession. Things appear to vary with the distance $x$ in a cyclic way, the same situation recurring again and again. Indeed, if the distance between successive wave crests is $\lambda$ (we call this the wavelength), then this is just the distance in which everything repeats. At a guess, I am going to add to the velocity potential a term like† :

$$\cos 2\pi x/\lambda.$$

You can see that if we increase $x$ by a distance equal to $\lambda$, then the angle increases by $2\pi$, which is a whole cycle, and the cosine returns to its original value.

Some people abbreviate this to :

$$\cos kx,$$

and call $k$ the 'wave-number'. You can see that the wave-number $k$ is just the same as $2\pi/\lambda$.

I do not yet know how much of this cosine term I ought to include so I will put in a numerical factor $C$ and write the velocity potential as :

$$\phi = cx + C\cos kx.$$

† In this formula the 'angle' $2\pi x/\lambda$ is to be understood as being radians, not degrees. We do this so that it is easy to differentiate the formula, as we do later.

The idea of this formula comes, as you know, from picturing a point that moves round and round a circle. Then the shadow or projection of it on some line moves to and fro about a mean position. But in writing this formula I am not suggesting that anything moves in a circle. It is only a mathematical trick to suggest one way in which $\phi$ might continually change its value in a repetitive way as we look farther and farther in the $x$ direction.

But the formula still won't do. I know that the water moves up and down too, so the velocity potential must have a gradient in the $z$ direction. I must somehow introduce $z$ into the formula.

This is where we make use of the rule that ensures that the water is not contracting or expanding. Because I do not think that $y$ need come into the formula for $\phi$, I can omit the middle one of the three terms and write the rule as:

$$d^2\phi/dx^2 + d^2\phi/dz^2 = 0.$$

I must include the $z$ in such a way that this rule is obeyed†.

† *The separation of variables*

I think you may be interested in an ingenious mathematical trick for finding formulae that satisfy this rule. A mathematician starts by supposing that the formula he wants, happens to be one in which the $x$ and the $z$ appear in separate factors. There is one expression that includes $x$ in some way, but not $z$, and another expression that includes $z$ and not $x$. I will denote them briefly as $X$ and $Z$. For instance, the expression $X$ might be the formula for that we have already thought of, while $Z$ depends on $z$ in a way we have yet to discover. The mathematician's assumption, then, is that

$$\phi = XZ.$$

Formulae are not necessarily like this of course. The formula $(x+z)$ looks simple enough, but can't be written in this way. Nevertheless, the mathematician's idea often works.

This formula, whatever it is, has to obey the rule:

$$d^2\phi/dx^2 + d^2\phi/dz^2 = 0.$$

In the first term, $\phi$ has to be differentiated twice with respect to $x$, while $z$ is treated as constant. This means that the expression $Z$ is not altered at all. The expression $X$ has to be differentiated twice. Without knowing just what $X$ is I can't say what $X$ becomes, but I will just represent the result as $X''$, with two primes to remind me that it is $X$ differentiated twice.

Then the first term is just $X''Z$.

Similarly the second term, where we differentiate twice for $z$, becomes $XZ''$.

So the expressions $X$ and $Z$ must obey the rule:

$$X''Z + Z''X = 0.$$

You can write this a little differently so that it reads:

$$X''/X = -Z''/Z.$$

Now comes the part of the argument that seems to me very ingenious. The expression on the left involves $x$ and you might think that its value would vary as you changed $x$. But this can't be so. It is always equal to the expression on the right that has no $x$ in it.

Equally well, the expression on the right contains $z$ but can't vary with $z$, being equal to the expression on the left that has no $z$ in it. Both expressions must be equal to some number, say $n$, that doesn't depend on either $x$ or $z$. So the $X$ and $Z$ must obey the rules:

$$X'' = Xn, \quad Z'' = -Zn,$$

though we don't yet know what the number $n$ should be.

*Putting z in the formula*

The first term in my velocity potential will do as it is. It describes a uniform stream with no compressions or expansions and I need no $z$ in it.

The $z$ expression should be attached to the second term that describes the waves, so I will use a mathematician's trick called the 'separation of variables' and guess that the second term is $ZC \cos kx$, where $Z$ is some formula involving $z$. I look first at the factor that has $x$ in it. Just to give it a name I call it $X$ and it is:

$$X = C \cos kx.$$

On differentiating this I find that

$$dX/dx = -kC \sin kx,$$

and by differentiating again I get the result:

$$X''(= d^2 X/dx^2) = -k^2 C \cos kx.$$

So it seems that the number $n$, which is $X''/X$, is $-k^2$.

Now we can find what $Z$ must be. It must obey the rule:

$$Z'' = k^2 Z.$$

What kind of function can $Z$ be? It can't be a sine or a cosine because this would always produce a negative sign when differentiated twice, and the $k^2$ is a positive number.

There are functions that give positive numbers. For instance $10^z$ or $2^z$ or $(0.5)^z$ are all functions of the variable $z$ and behave in the way we want†. But if you have met the 'exponential' function you

---

† They are all some constant positive number raised to the *power z*. They are 'exponential' functions of $z$. If you have not met the idea I will try to show how it meets our needs.

If I could find some expression (some 'function' of $z$, and I will continue to call it $Z$ for short) in which the gradient $dZ/dz$ was proportional to the function itself, I would know that

$$dZ/dz = kZ.$$

I have assumed that the proportionality factor $k$ is a constant so I can differentiate again and find that

$$d^2 Z/dz^2 = k\, dZ/dz = k^2 Z.$$

It would not matter whether $k$ was positive or negative because $k^2$ would be positive in either case, so this is the sort of function I want.

What would this function $Z$ be like? How ought it to depend on $z$? We could try to build up a graph. I will plot the variable $z$ horizontally and draw the magnitude of the function $Z$ vertically in the usual way. In the wave problem, of course, $z$ itself represents vertical height but I am not drawing the waves now but merely trying to draw a picture of the unknown function $Z$. I will start off at the point $z = 0$. I do not know what the value of the function $Z$ might be at this place. I will call it unity and see what happens.

The slope of the curve is to be proportional to the size of the function, the ordinate of the curve. For the sake of drawing a curve I will say that the constant of proportionality is positive. So the curve begins with an upward slope. The slope increases as I extend the curve farther, because it is proportional to the ordinate and this is increasing. I sketch a curve like this as far as $z = 1$. It is largely guesswork and I don't know just how big the ordinate will be when I reach $z = 1$, but I will suppose it has reached a value $V$.

When I begin to sketch the curve between $z = 1$ and $z = 2$ it strikes me that the shape should be very much the same. The slope at $z = 1$ is bigger than it was at $z = 0$ but this is because the ordinate is bigger. Imagine that

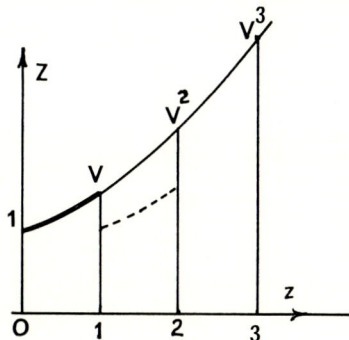

I tried to sketch this part at a reduced scale. I would need to make the slope smaller in proportion. If at $z = 1$, I chose the reduced scale so that the ordinate was as high as it was originally at $z = 0$, then the slope would need to be just the same as it was at $z = 0$. On this reduced scale the curve between $z = 1$ and $z = 2$ would exactly duplicate the previous part.

So when I use the same scale throughout I see that the ordinate increases by the same factor in the two steps. At $z = 0$ the ordinate is 1, at $z = 1$ the ordinate is $V$, so at $z = 2$ the ordinate must be greater again by a factor $V$, that is to say $V^2$, and by going to $z = 3$, I would produce another factor of $V$ and the ordinate becomes $V^3$.

Indeed I see that the exponent of $V$ (the power to which it is raised) is just the value of the variable $z$. The proper equation for my curve is $V^z$. It is called an 'exponential' function of $z$.

I can choose the value of $V$ to suit my needs. If $V$ is only a little larger than 1 the curve rises only slowly. If $V$ is large the curve rises steeply. If $V$ is chosen less than 1 (but positive) then the curve slopes downward and the ordinate gets less and less though it never goes to zero exactly. I would not try making $V$ negative because I would not know what the formula meant in that case.

Mathematicians write the formula a little differently. You can see that this is possible ; $(\sqrt{V})^2$ means just the same thing as $V$ so I could write $V^z$ as :

$$(\sqrt{V})^{2z},$$

and it is the same thing, though there is now a factor of 2 in the exponent. I spoke of taking different values for $V$, but mathematicians prefer always to use the same number, which they call $e$ and which is actually an 'irrational' number that I can't express by fractions or decimals but is near to 2·718.... They give the curve different shapes by altering the factor in the exponent. For instance $e^{3z}$ gives a steep graph, $e^{0·5z}$ gives a flatter one and $e^{-0·5z}$ gives a curve that slopes downwards.

will know that it is more usual to write the formula for the function $Z$ in the form $\exp(kz)$, which is short for $e^{kz}$ or $(2 \cdot 718..)^{kz}$. One chooses the constant number $k$ to suit the problem. The rule for differentiating this function is (though I won't try to prove it here):

$$\mathrm{d}[\exp(kz)]/\mathrm{d}z = k\exp(kz).$$

Doing this again, we find by the same rule, that

$$\mathrm{d}^2[\exp(kz)]/\mathrm{d}z^2 = k\mathrm{d}[\exp(kz)]/\mathrm{d}z = k^2\exp(kz).$$

Evidently the formula $\exp(kz)$ does exactly what we want the function $Z$ to do. It gives the positive factor $k^2$ when differentiated twice.

Oddly enough you get the same number $k^2$ if you adopt a formula $\exp(-kz)$. You can see that $(-k)$ times $(-k)$ is still $k^2$. So I could write this function $Z$, that I need to have in the velocity potential, either as:

$$\exp(kz)$$

or I could put it in as:

$$\exp(-kz).$$

For the time being I will choose the first alternative and see what happens.

*The wave formula*

Our guess at a formula has now become:

$$\phi = cx + C\exp(kz)\cos kx.$$

It describes a steady backward stream (the first term) and the second term puts some cyclic disturbance into the stream.

We don't yet know that the second term describes the right kind of disturbance. Let's wait and see.

There is a further condition that must be fulfilled. The waves on this moving stream have a free surface exposed to the atmosphere. The air does not drag on the water (we suppose) but it will exert a pressure. The pressure on the moving water everywhere over the surface must be equal to the atmospheric pressure. The actual value of the atmospheric pressure is not important. The important feature is that it must be the same everywhere.

We must therefore check that this formula implies a constant pressure in the fluid at the free surface. To do this we must find an expression for the pressure in a moving fluid.

*The formula for pressure in steady-state flow*

In our problem the surface itself is a stream line, so I will find how pressure must vary along any stream line in a steady flow.

A cluster of stream lines can be pictured as a tube with the water flowing steadily through it, and it is easy to find the pressure by an energy argument.

Think of the water that fills this tube between positions 1 and 2. The walls of this tube are nothing more, of course, than the surrounding water. Nevertheless, the water we are thinking of behaves as if it were flowing down a frictionless tube.

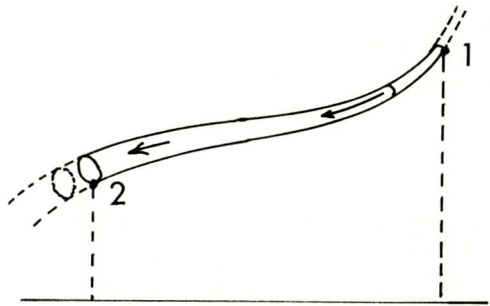

Fig. 3.4. A tube of stream lines in steady state flow.

Think of what happens as the water moves forward during some very short time, say $\Delta t$. At position 1 the water that follows is pressing forward and doing work on the piece of water we are thinking of. The work done is:

pressure × area × distance moved.

If at position 1, I call the pressure $p_1$, the cross-section $A_1$ and the forward speed $q_1$, I think you will agree that the work done is:

$$p_1 A_1 q_1 \Delta t.$$

But at position 2, the piece of water we are thinking of is doing work on the water ahead as it pushes forward. So far as our piece of water is concerned, this is energy lost. If I use the subscript 2 for the pressure etc. at position 2, I think you will agree that the overall energy gained by our piece of water is:

$$\text{gain in energy} = p_1 A_1 q_1 \Delta t - p_2 A_2 q_2 \Delta t.$$

This increase in energy must show itself as a gain in kinetic energy or gravitational potential energy. Now, it is true that every small piece of water is probably changing its speed and changing its level as it moves. It would be difficult to argue about all the various small pieces if we did not remember that this is a steady flow. As each small piece of water moves forward to a new position it acquires the same position and speed (and therefore the same energy) as the piece of water that was there before. When the column of water between positions 1 and 2 moves forward, the change in energy amounts merely to this, that a certain mass of water, say $M$, is missing from position 1 and an equal mass has appeared at position 2 but with different kinetic and potential energies.

If I use $z$ to denote distances measured vertically upwards then a mass $M$ has, in effect, been raised a distance $(z_2-z_1)$. It has also increased its speed from $q_1$ to $q_2$. The overall gain in kinetic and potential energy amounts to:

$$\text{gain in energy} = Mg(z_2-z_1) + \tfrac{1}{2}M(q_2^2-q_1^2).$$

Looking back at the first equation you will see that

$$A_1 q_1 \Delta t$$

is just the volume that is missing at position 1, while

$$A_2 q_2 \Delta t$$

is just the volume that has appeared at position 2. If $\rho_1$ and $\rho_2$ are the densities of the fluid at these two places, then these volumes are just $M/\rho_1$ and $M/\rho_2$, so the first equation for the gain in energy could be written:

$$\text{gain in energy} = M(p_1/\rho_1 - p_2/\rho_2).$$

When you equate these two expressions for the gain in energy, it shows how the pressure varies with the level and the speed. You can write it (after cancelling $M$ throughout):

$$p_1/\rho_1 + gz_1 + \tfrac{1}{2}q_1^2 = p_2/\rho_2 + gz_2 + \tfrac{1}{2}q_2^2.$$

It shows that the quantity

$$p/\rho + gz + \tfrac{1}{2}q^2$$

remains constant along a stream line in steady flow.

We have proved this even if the density should vary, as it could in a compressible fluid like air. However, we are concerned with water waves and shall suppose that the density does not change.

I think you will agree that the result we have found looks reasonable. It says that if the speed $q$ gets less, then this will tend to mean a greater pressure. Quite so; it is the extra pressure ahead of the moving fluid that causes the fluid to decrease its speed. The term including $z$ shows the effect of gravity. If the fluid moves downward, $z$ decreasing, then it moves to places where the pressure is greater, as one would expect.

*Is our wave formula right?*

We thought of a formula for velocity potential:

$$\phi = cx + C \exp(kz) \cos kx,$$

and the question now is whether this formula describes a motion in which the pressure at the free surface is constant.

We shall find this difficult to justify, exactly.

The reason is, that it isn't true, exactly. You have met at least one situation like this before. Has it not struck you that the proof of the period of a simple pendulum is rather messy and inexact? You try to show that the period is given by:

$$T = 2\pi \sqrt{(l/g)},$$

and you find it difficult to give a clean exact argument. This is because it isn't true, exactly.

Yet the formula for the period of a pendulum is nearly true in the following sense. If you make experiments with a pendulum swinging 90° on either side of the vertical, you find that the period is actually about 18% greater than the formula predicts. If you reduce the

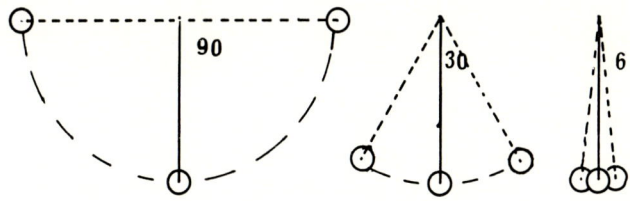

Fig. 3.5.

angle to 30°, then the actual period is only 2% greater than the value given by the formula. If you let the pendulum swing only 6° on either side, and if you could do the experiment precisely enough, you would find that the formula was wrong by only 0·06%.

Our proposed formula for a wave-train is wrong, or right, in a similar kind of way. It works out quite well enough so long as the waves that we imagine are not too steep, so long as the slopes of the surface are small.

I will try to show this. I must first see what the waves of our formula look like.

*The profile of the waves*

First we can deduce the velocities. These are:

horizontal: $u = -d\phi/dx = -c + kC \exp(kz) \sin kx,$

vertical: $w = -d\phi/dz = \quad -kC \exp(kz) \cos kx.$

These show the velocities at any position $x$ and $z$. Now I want to find where the particles go to.

You can see that if there were no waves at all, which would be the situation if I took the number $C$ to be zero, then the horizontal velocity would be the same everywhere and equal to $-c$ while the vertical velocity would be zero.

It is easy to see where each particle goes in that case. If I start with a particle at position $x = 0$, $z = 0$ at time $t = 0$, then later on after a time $t$ its position is just:

$$x = -ct, \quad z = 0.$$

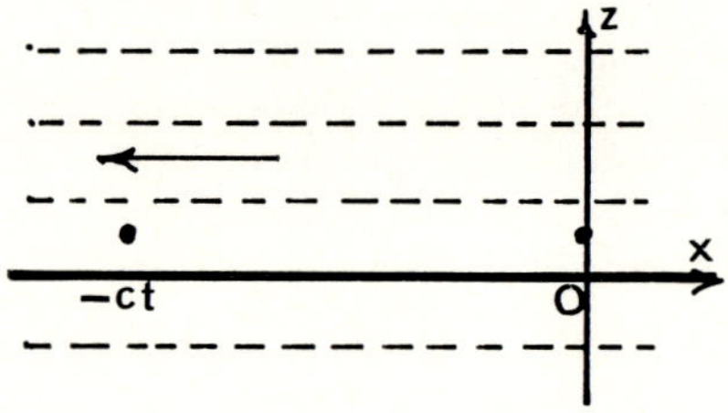

Fig. 3.6. The approximate path of a particle.

I now imagine that some waves are present, and the number $C$ is not zero. The previous formulae for $u$ and $w$ show the velocities a particle has, so long as I know where it is. Well, I just now picked on a particle and found where it went, getting the positions $-ct$ and $0$ as above. I might substitute these in the formulae and find the velocities of this particle at that instant. Doing this I get:

$$u = -c - kC \sin kct,$$
$$w = \quad -kC \cos kct.$$

You can see this must be wrong. When waves are present the particle will not be quite at the position I said it was. Its velocities will be a little different in consequence.

But this is where I must approximate. If the wave disturbance is small it won't move the particle far, and these formulae may be good enough. Let's go on and see what happens.

What formulae for position would suit these formulae for the velocities? I think you will approve if I guess the following answer:

$$x = -ct + \frac{C}{c} \cos kct,$$

$$z = \quad -\frac{C}{c} \sin kct.$$

You can see that if I differentiate these formulae, the formula for $dx/dt$ is just the same as that for $u$, above, and the formula for $dz/dt$

is just the same as that for $w$, which is what I want. I think these formulae for $x$ and $z$ will do†.

What kind of motion do these formulae describe? In the formula for $x$, the term $-ct$ means a continual progress down-stream. This is the idea we began with, but now there is an extra term showing an

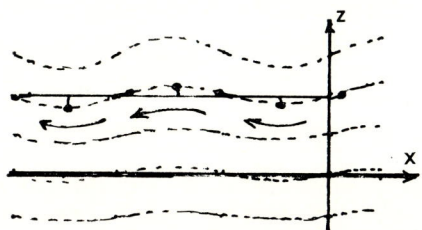

Fig. 3.7. A better approximation to the path of a particle. To this approximation the waves add a small circular motion to the steady down-stream flow.

additional oscillation forwards and backwards. In the formula for $z$ it seems that the particle does not stay exactly at the level $z = 0$ but oscillates above and below this level. The amplitude of both these oscillations is $C/c$ and perhaps you may agree, because one is a cosine term and the other is a sine term, that the actual motion amounts to a small circular motion added on to the down-stream drift.

Now we can think about the pressure and see whether it is the same all over the free surface.

*The condition for constant pressure on the free surface*

What picture have we built up?

There is a stream flowing in the negative $x$ direction with a speed $c$.

On this stream there are waves that are 'stationary' in the sense that they stay in the same place always. They stay in the same

---

† Perhaps at this stage you may complain that the formulae could be different. I could have included some constant term in both $x$ and in $z$, and yet on differentiating I would still have found the same values for $dx/dt$ and $dz/dt$. The question is whether or not I need to include any constant terms.

Remember that originally I thought of the particle drifting down the stream, and I chose to think of a particle that happened to pass through the position $x = 0$, $z = 0$ at the instant $t = 0$. You can see that the formulae for $x$ and $z$ do give this result so long as I suppose that there are no waves ($C$ is zero).

On the other hand, if waves are present, so that $C$ is not zero, the guessed formula for $x$ at the instant $t = 0$ does not give zero, it gives $x = C/c$. But perhaps this is reasonable. As the particle moves down the stream, the waves move it about, both up and down and to and fro relative to the position that the steady stream alone would put it in. When waves are present it is perhaps reasonable that the particle should not be exactly at $x = 0$ at the instant $t = 0$.

I prefer therefore not to include any constant terms in the formulae for $x$ and $z$.

position as the stream passes through them. I called it a 'steady-state' situation.

If a particle of water is on the surface of such a stream it will stay on the surface, carried along by the stream and going down into the wave troughs and rising over the wave crests as it passes them. The surface of such a 'steady-state' stream is therefore a stream line. We have already found how to trace the variation of pressure along a stream line in a 'steady-state' flow.

The shape of the stream surface is given by the formulae we have already found for the path of a water particle. When I chose to think of a particle that was approximately at the level† $z = 0$, I did not say that it was a particle on the surface; but it could have been.

I can choose to have the surface of the stream at any level I wish, but merely for reasons of simplicity I propose that it shall be at about $z = 0$. I propose to follow a particle as it moves along the surface stream line and to discover whether the fluid pressure at particle stays constant, as it should. We have already found formulae for the velocities and co-ordinate positions of such a particle, and to quote them again they are:

$$u = -c - kC \sin kct,$$

$$w = \phantom{-c} -kC \cos kct,$$

$$x = -ct + \frac{C}{c} \cos kct,$$

$$z = \phantom{-ct} -\frac{C}{c} \sin kct.$$

† My real reason for choosing $z = 0$ was merely that it led to more simple-looking formulae. The factor $\exp(kz)$ did not need to appear continually. The whole argument could have been made for a particle that was approximately at the level $z = h$ (whatever the height $h$ might be) and then we would have found that the formulae for the particle's position were:

$$x = -ct + \frac{C}{c} \exp(kh) \cos kct,$$

$$z = \phantom{-ct} -\frac{C}{c} \exp(kh) \sin kct.$$

I hope you agree that this is so.

Just as an aside, I would like to point out that the formula now shows that the amount of wave disturbance is different at different levels. The amplitude of the disturbance due to the waves is $C/c \exp(kh)$ in both the horizontal and vertical directions. If one takes a high position, $h$ positive, the disturbance is large because the factor $\exp(kh)$ is large. If one looks at a lower place in the stream, $h$ negative, the disturbance is small because the factor $\exp(kh)$ is small. The factor happens to be unity at the level $h = 0$ and then the amplitude is just $C/c$ as we found before.

The particle we are thinking of moves along a stream line and we know that the pressure, elevation and speed are connected by the rule :

$$p/\rho + gz + \tfrac{1}{2}q^2 = \text{constant (independent of time)}.$$

Let us substitute for the elevation $z$ and for $q^2$ (which is just $u^2+w^2$). We get, if you work it out :

$$p/\rho - g\frac{C}{c}\sin kct + \tfrac{1}{2}(c^2 + k^2 C^2) + ckC\sin kct = \text{constant}.$$

The pressure in this equation can be independent of time only if the terms in $\sin kct$ happen to cancel.

This requires that

$$g\frac{C}{c} = ckC,$$

which gives :

$$c^2 = g/k.$$

The pressure at the particle will be constant if this condition is obeyed. It is a formula for the speed required in the stream. It depends on the wave-number, and consequently on the wavelength, of the waves. The surface pressure in the fluid will be constant only if the stream moves at just this speed.

*The speed of a wave-train*

Now, for a moment, let's return to the original point of view where we stay with the water and see it more or less at rest while the waves move forward through it.

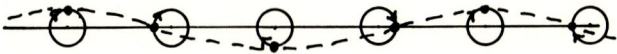

Fig. 3.8. The approximate particle motion when we watch the waves go past.

Instead of seeing the stream move backward at a speed $c$, we see the waves move forward at a speed $c$. The formula we have just found :

$$c^2 = g/k,$$

tells us the speed of travel of the wave-train.

What about the motion of the water ?

In the steady state we saw that a particle travelled down-stream with speed $c$ but that in addition it had a small circular motion. Now only the circular motion is left. When we have no stream and the wave-train travels forward through the water, the water particles move in circles.

It is easy to write the formula. We found the positions $x, z$ for a particle moving down-stream in the ' steady state '. All we need to do is to cancel the term $-ct$ in the formula for $x$ and look at what is left. It is:

$$x = \frac{C}{c} \cos kct,$$

$$z = -\frac{C}{c} \sin kct.$$

What does the particle do? Start with the instant $t = 0$; then $z$ is zero but $x$ is $C/c$. The particle is displaced in the direction of wave travel.

Now think of a later time when the phase angle $kct$ has increased by a quarter cycle (is $\pi/2$ in radian measure). Then $x$ is zero but $z$ is $-C/c$. The particle has moved backwards and sunk. It is in the trough of a wave.

You can see that it continues to move backwards, because when the phase angle has reached a half cycle, or $\pi$ in radian measure, the value of $x$ is $-C/c$. Also the particle is beginning to rise in the next wave because $z$ has become zero again.

Finally, after another quarter cycle, which brings the cosine to zero and the sine to $-1$, the particle is on the top of a wave because $z$ is $C/c$, and it is moving forwards towards its original position.

The particle moves in a vertical circle, going forward with the waves when it is in a wave crest and moving backwards when it is in a trough.

*This formula only suits a wave in deep water*

I began by following a particle that started at the level $z = 0$, so the factor $\exp(kz)$ in the formula disappeared, being just 1. But I need not have done this. If I had started at a level $z$ I would have found a factor $\exp(kz)$ in all the equations. For instance, the vertical velocity $w$ would have been:

$$w = kC \exp(kz) \cos kct.$$

This factor shows us that the wave disturbance is different at different levels. If you go downwards, making $z$ negative, the factor $\exp(kz)$ becomes small. Wave activity fades away as you go down below the sea surface.

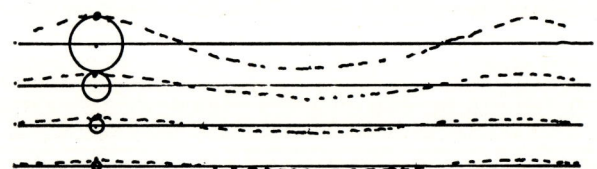

Fig. 3.9. Our formula gives a particle motion that dies away with increasing distance below the surface but never becomes quite zero. The formula suits a wave over deep water.

This fact is well known to people in submarines. When dinner-time comes, they submerge the vessel and eat dinner in peace and comfort under water.

For example, rough waves at sea may have a wavelength, $\lambda$, of about 100 metre. What happens if one goes down a distance of 20 metre below the surface? The factor $\exp(kz)$ shows that the wave activity is reduced by a factor:

$$[\exp(-2\pi \times 20/100)] = 0.29 \quad \text{(from tables)}.$$

The disturbances due to the waves are less than a third of what they are at the surface.

But the factor never falls quite to zero. There is always some up-and-down motion. So the formula we have invented is not suitable for picturing waves in water over a sea-bed because we cannot have water moving up and down through the sea-bed. Our formulae, and in particular

$$c^2 = g/k,$$

refer to a train of waves in quite deep water.

You can work out the pressure on some particle below the surface if you wish. You need to have an $\exp(kz)$ in the formulae. The pressure will be greater there than it is on the surface, of course, but you will still find that it does not fluctuate with time. This seems curious. It is true, however, only for these deep-water waves.

When waves pass over a flat sea-bed, the water motion is rather different as we shall see in the next section. Then you could use the pressure formula to show that the pressure at a particle on the sea-bed does fluctuate as the waves pass. Indeed, instruments for measuring sea waves usually stand on the sea-bed and detect the fluctuating pressure.

## A wave-train over a flat sea-bed

How can we make up a formula to describe this? There needs to be a certain level (the level of the sea-bed) at which the water doesn't move up and down at all. Think again of the 'steady-state' picture, where the water runs past but the waves are stationary.

You will recall that when we were trying to include $z$ in the wave formula it turned out that there were two possibilities. I adopted the factor $\exp(kz)$. The alternative was a factor $\exp(-kz)$ and I will use that now and write:

$$\phi = cx + C\exp(-kz)\cos kx.$$

This formula is good in many ways. It gives irrotational motion (of course, because it is a velocity potential). It doesn't require the water to expand or contract. It pictures a steady backward stream with a cyclic disturbance on it.

But there is a factor $\exp(-kz)$. You can see that the wave fades out as you go *upwards*. It is an upside-down wave. It is not much good to us as it stands†.

It happens, however, that this formula gives vertical velocities that are in the opposite sense to those given by the formula we used first. The reason is that when you differentiate $\exp(-kz)$ you get a minus sign. But the horizontal velocities in both formulae are in the same sense.

So, you see, it should be possible to combine the two formulae and have the vertical velocities cancel at some level. Then this could be the level of a flat sea-bed. Let's try. I will write as a trial:

$$\phi = cx + C[\exp(kz) + \exp(-kz)] \cos kx.$$

Now I differentiate to find the velocities of the water:

$$u = -c + kC[\exp(kz) + \exp(-kz)] \sin kx,$$
$$w = -kC[\exp(kz) - \exp(-kz)] \cos kx.$$

You can see now that if I choose the level $z = 0$, the $\exp(kz)$ and $\exp(-kz)$ are each equal to 1 and the formula for the vertical velocity $w$ gives just zero. The level $z = 0$ will have to be my sea-bed.

*The motion of a particle*

I will think of a particle at some level $z = h$. It will remain approximately at this level, rising and falling in the waves as it runs with the stream. For its position at time $t$ I could write, approximately only:

$$x = -ct, \quad z = h.$$

I put these in the formulae for the velocities and find:

$$u = -c - kC[\exp(kh) + \exp(-kh)] \sin kct,$$
$$w = -kC[\exp(kh) - \exp(-kh)] \cos kct.$$

Now I have to pick the right formulae for the position of the particle. I think that they should be:

$$x = -ct + \frac{C}{c}[\exp(kh) + \exp(-kh)] \cos kct,$$
$$z = h - \frac{C}{c}[\exp(kh) - \exp(-kh)] \sin kct.$$

These are a better estimate of the particle's position‡.

† You actually can have waves of this kind. When waves travel over water, the air above is pushed out of the way. It turns out that the air over the water moves in this pattern of an upside-down wave.

‡ How has the first term, $h$, come into the $z$ equation? Integration of a velocity gives the overall *change* in position, not the actual position. One is free to add on any constant value. I have added the $h$, and now you see that if the waves were very weak ($C = 0$) the particle would just be at $x = -ct$, $z = h$ as I supposed at first.

The first terms show how the particle runs backward with the stream. The second terms show the additional oscillations due to the waves. I can see that if I looked at the level near the sea-bed ($h$ quite small) then the up-and-down oscillations would tend to zero because $\exp(kh)$ and $\exp(-kh)$ both tend to the value 1, and cancel. This is what I wanted.

But the horizontal oscillations, those in $x$, would not go to zero. Near the sea-bed, the waves cause horizontal oscillations, but the vertical oscillations are small or zero. This looks reasonable.

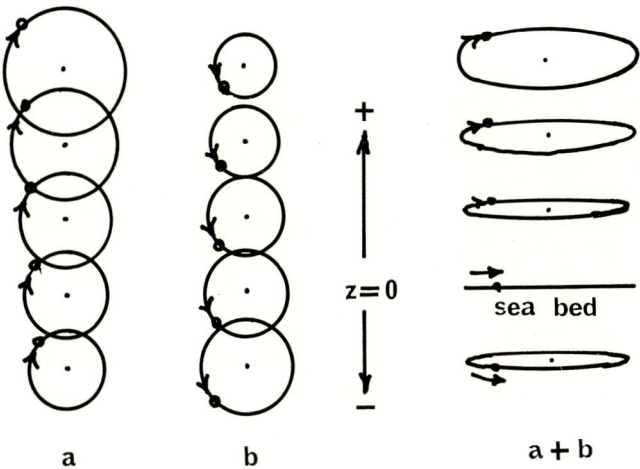

Fig. 3.10. By superposing the motions in a deep-water wave and in an 'upside down' wave we get a motion that suits waves over a flat sea-bed (at $z = 0$).

## Constant pressure on the surface

If the particle I am thinking of is on the surface, then $h$ is the depth of the sea.

I must make sure the surface particle stays at a constant pressure. As before I take the formula that holds for a stream line:

$$p/\rho + gz + \tfrac{1}{2}(u^2 + w^2) = \text{constant},$$

and substitute the values of $z$, $u$ and $w$ that I have found†.

I find on doing this that the speed of the stream must be such that

$$c^2 = Ag/Bk,$$

where $A$ is the vertical amplitude and $B$ is the horizontal amplitude of the wave oscillations at the surface.

† I won't write it all down, but leave you to do so. Notice that it is fair in our approximation to ignore terms in $C^2$.

Finally we imagine ourselves staying with the water and letting the waves move past us. Then of course we see the waves travelling forward with this speed $c$ that the formula gives.

Curiously enough this is exactly the formula we guessed in Chapter 1. But then we could not guess what the ratio $A/B$ should be. Now we see that it is:

$$A/B = [\exp(kh) - \exp(-kh)]/[\exp(kh) + \exp(-kh)].$$

If you wished to, you could work out the value of this ratio for any particular kind of wave. The depth of water is $h$, and $k$ is $2\pi/$(wavelength). Indeed, you could even look it up in mathematical tables because this ratio has a name. It is called:

$$\tanh kh.$$

It is always less than 1.

Now you see that, when there is a flat sea-bed, the water particles do not move in circles. They move on oval (elliptical) paths that are flatter and flatter as you look nearer the sea-bed. I have pictured this in fig. 3.10.

Again, if $A/B$ at the surface is less than 1 it makes the wave speed less. Waves in shallow water travel less rapidly than do waves of the same wavelength in deep water.

In ' very shallow ' water, where $kh$ is very small, you can perhaps show that the ratio of $A/B$ tends to $kh$. Then the formula for wave speed becomes just:

$$c^2 = gh,$$

and all waves travel at the same speed, as we guessed in Chapter 1.

*The end of the mathematics*

In the last two chapters I have tried to give a genuine account of how people are able to reason about fluid motions. It has been mathematical and you may have liked it or you may not. But this is the end. From now on we can think almost entirely in mental pictures.

I shall take trains of waves, perhaps with different wavelengths, perhaps travelling in different directions and add their effects. I add together trains of straight waves and find that I can imagine more complicated things, such as standing waves in a swimming-pool, circular waves in a teacup, waves spreading from storms at sea, and the complicated pattern of waves that follow a ship. I expect you will find that these arguments are the sort you could have made up for yourself if you had needed to. It just happened that you had no reason to think of them. Physics is all like that. Every argument that you meet in physics was made up by someone, made up just as you make up arguments yourself. They didn't *learn* it, they made it up.

Consequently, the right way of looking at physics is to keep asking yourself : How could I myself have invented that idea, supposing no one had told me of it ?

It is the right way to look at mathematics, too.

# CHAPTER 4
## making up wave patterns

HERE are some pictures of our perfect wave-train.

Figure 4.1 shows what happens in very deep water. All the water particles move in circles. The circles are smaller at greater depth because the wave motion dies away with depth.

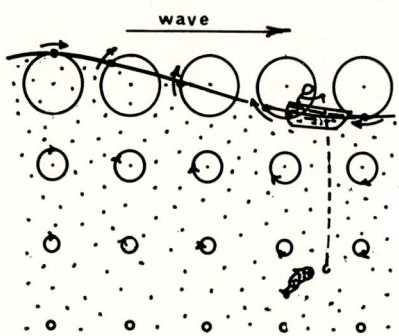

Fig. 4.1. A wave in deep water.

The surface particles don't move in unison. One particle reaches the top of its ' orbit ' and begins to sink. Then a particle farther along repeats this, and then the same thing happens still farther along the line of progress. This is how the effect of a travelling pattern, a ' wave ', is created.

I have had space to show the details of only one crest and one trough, but you can see how the diagram could be extended to represent a whole succession of waves.

I have drawn a small boat in the trough. The wave crest on the left is higher than the boat. Do you think that as the wave crest travels along, it will swamp the boat ?

I don't think that it would. When the wave crest reaches the boat, this merely means that the water under the boat has risen as far as it will. As the water rises it carries the boat up with it.

Figure 4.2 pictures a wave where there is a flat sea-bed not far below the surface. The orbits of the particles are long ovals now. Below the surface the ovals are flatter until at the sea-bed the water merely moves to and fro in a horizontal line.

*Adding wave-trains together*

I shall try to build up more complicated wave pictures by adding wave-trains together.

That is to say, I shall take the motion that a water particle would have in just one of the wave-trains and add it to the motion that the other wave-train by itself would give it. I shall say that the combination is just what the particle would do if both wave-trains happened to arrive at the same time. This is usually called 'superposing' two wave-trains.

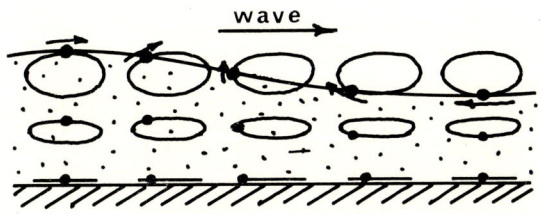

Fig. 4.2. A wave in water over a flat bed.

This method of superposing wave-trains is good enough so long as the waves are 'low'. We met a similar difficulty with the wave-trains themselves. Our pictures of the wave-trains are not quite correct but they will do very well so long as we do not think of waves with steep slopes. The same is true of our combined wave-trains.

When I add wave-trains my own inclination is to think in pictures and imagine how things work out. But sometimes it is convenient to check one's reasoning by algebra. Here are some formulae that I will use if I want to think in that way.

If I wish to say mathematically that the water surface at some place is rising and falling I will write:

$$\text{elevation} = a \cos(\omega t - \alpha).$$

I put in a suitable number $\alpha$ to make the top position occur when I wish. You can see that the top position (zero angle) comes at the moment $t = \alpha/\omega$.

If I wish to describe a wave profile moving along, I will write:

$$\text{elevation} = a \cos(\omega t - kx - \alpha).$$

The $\alpha$ just alters the timing; I may not need it. The term in $x$ makes the elevation vary 'sinusoidally' in distance. You can see from the negative sign that if I want to keep pace with the wave (keep the angle the same), then as time increases I must increase the $x$ distance. So this wave is moving forward in the $+x$ direction. On the other hand, if I write the $kx$ with the same sign as the $\omega t$:

$$\text{elevation} = a \cos(\omega t + kx),$$

this would mean a wave moving in the $-x$ direction; I would need to reduce $x$ as $t$ increased in order to keep looking at the same place on the wave.

*Similar waves going in the same direction*

These always add up to just another wave-train.

You might wish to add things up mathematically, and perhaps write:

$$\text{elevation} = a\cos(\omega t - kx - \alpha) + b\cos(\omega t - kx - \beta),$$

but I leave it to you to work that one out.

I prefer to think of a diagram. You know that these cosine quantities can be thought of as the ' projection ' of a rotating arm. If I take an arm of length $a$ at an angle $(\omega t - kx - \alpha)$ the projection gives me the first term above. If I want to add another cosine quantity, I can choose a second arm of the right length and at the proper angle and add it to the end of the first one.

Now in this case the two angles change with time, or with $x$, in exactly the same way (since $\omega$ and $k$ are the same in both). The two arms might just as well be connected rigidly or even replaced by a single arm.

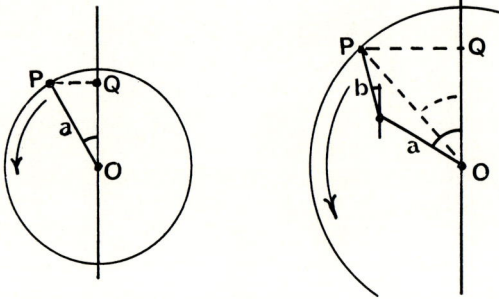

Fig. 4.3. One way of arguing about the way in which two ' simple harmonic motions ' combine.

So the combination is just another wave-train like the ones I started with.

I find this a useful picture to have in mind. It is so easy to see what the result will be.

*Equal wave-trains going in opposite directions*

Figure 4.4 is my attempt to picture what happens.

On the left I have sketched the two waves. The full lines represent the wave-train that travels to the right. The broken lines represent the wave-train that travels to the left. I have drawn the wavy outline of the water surface in each wave-train and have put arrows to show the direction of the water motion at different places.

On the right I have pictured the combined wave. I have added everywhere the elevations given by the separate waves, and this gives me the profile of the combined wave. I have also combined the velocities, vectorially, to show the wave velocity in the combined wave.

The successive pictures show successive stages as time passes.

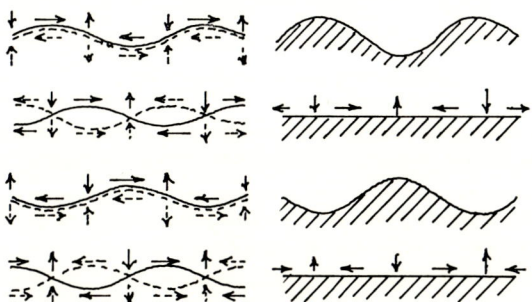

Fig. 4.4. The history of a standing wave.

In the first stage the wave crests of the separate trains happen to coincide in position, and so do the wave troughs, so the combination is a wave pattern twice as high. But, curiously enough, you see that the water velocities in the separate wave-trains are everywhere just equal and opposite. They add up to zero everywhere; every particle of water in the combined wave is still, for the moment.

In the second stage I have pictured the state of affairs one-quarter of a cycle later. The separate wave-trains each have advanced one-quarter of a wavelength, one going to the right, the other to the left. Now you see that the crests of one wave-train coincide with the troughs of the other. The elevations given by the separate wave-trains add up to zero everywhere. If you photographed the combined wave at this stage, you might say that there was no wave motion at all. However, you can see from the diagrams that the water velocities that the trains give are now just the same everywhere. They add up everywhere, and this means that the water in the combined wave is rapidly moving. The surface is flat now, but where there used to be a wave crest, the water is moving downwards. Where there used to be a trough it is moving upwards, and at places in between it is moving more or less horizontally.

In the third stage, when the two wave-trains have advanced another quarter wavelength in opposite ways, you can see that the velocities cancel once more but that the elevations add. There is again a large combined wave and all the water particles are momentarily at rest. Notice, however, that the wave has turned 'inside out'. The original crests have become troughs, and the troughs have risen up to form crests.

In the fourth stage, after another quarter cycle, the elevations have disappeared again and all the water is moving rapidly. If you look at the arrows showing the directions in which the water is moving in the combined wave, you can see that these are the right directions for the water to move, so that shortly, after another quarter cycle, the water will show crests and troughs again as it did in the first stage of this series of sketches.

So the combined wave pattern does not travel. It does not go either to right or left. Nevertheless there are rapid changes going on, indeed the wave continually turns ' inside out ', and it is quite different from the ' stationary ' waves on a moving stream of water. I will call the combined wave a ' standing ' wave because it alternately stands up and sits down†.

You can sometimes find these standing waves when a nice travelling wave-train strikes a flat wall at right angles and is reflected back on itself. The waves need to be quite low to give the smooth outlines that I have used in fig. 4.4. If the waves are steep the crests tend to be more pointed, or ridge-like, and indeed you can have water shot up into the air. I have somewhat exaggerated the height in the diagrams; you should look for waves less steep than the diagrams suggest.

You can make an argument using formulae if you prefer. So far as the elevation is concerned, you can write the elevation due to one wave-train as:

elevation at position $x$, time $t = a \cos(\omega t - kx)$.

This represents the wave-train travelling to the right (the positive $x$ direction). For the wave-train going the other way:

elevation at position $x$, time $t = a \cos(\omega t + kx)$.

You will be able to show that the sum of these can be written:

elevation at $x$ and $t$ in the standing wave $= 2a \cos kx \cos \omega t$.

The term $\cos \omega t$ shows that the elevation changes, the surface rises and falls. However, the ' amplitude ' of this motion is $2a \cos kx$. At some places, say $x = 0$, it is largest and equal to $2a$. At other places, $\cos kx$ will be zero, showing that there is no rise and fall. At other places you find that the ' amplitude ' turns out to be $-2a$. The negative sign merely means that the deflection is in the reverse sense; it is a place where the water has sunk to make a trough at the moment

---

† Some books use the names ' stationary ' and ' standing ' as I have done and some use them in the reverse way. There seems to be no accepted convention, but I hope that I have made it clear how I intend to use the names in this book.

when water a little farther along has risen to make a crest. But half a cycle later, the wave would have turned ' inside out ' and the trough would be a crest.

These formulae refer only to the vertical displacement that I previously called $z$. You could write similar formulae to find what happens in the horizontal motion.

Just to complete the picture, imagine what a boat would do in a standing wave. If it happened to be just at a crest or trough it would go up and down. If it happened to be halfway between these places it would move from side to side. At other places it would move to and fro along a sloping line. I think that there would be some rolling of the boat too, for the water surface is tilting.

You could also use the ' rotating arm ' diagram to show how two travelling wave-trains can combine to make a standing wave. You might yourself try this.

*Two standing waves can make a progressive wave*

This is a curious idea but it can be useful in an argument.

You must choose the standing waves so that they have equal amplitudes. You must position them so that they are just one-quarter wavelength different in position, and then make them one-quarter cycle out of step in time. Thus, we saw just now a formula for a standing wave, it was :

$$\text{elevation} = a \cos kx \cos \omega t.$$

For the second standing wave I will write :

$$\text{elevation} = a \sin kx \sin \omega t,$$

for I think you will agree that putting sines instead of cosines makes this second wave just one-quarter cycle different in both position and time. The sum of these is just

$$\text{elevation} = a \cos(\omega t - kx),$$

which is a progressive wave of the same amplitude.

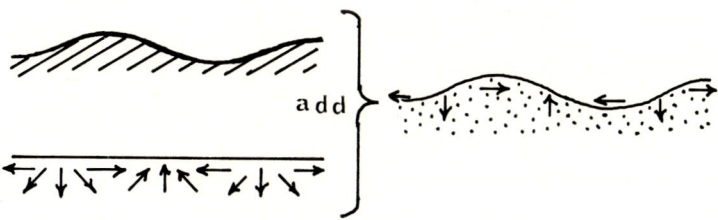

Fig. 4.5. A progressive wave can be thought of as two standing waves combined.

To think of it pictorially you should imagine the instant at which one of the waves has reached its maximum elevation. This presents a picture of lines of crest and trough, but so far as this wave is concerned the water is momentarily at rest.

But the second wave is just one-quarter cycle out of step in time. This wave will be in its 'flat' state. It adds nothing to the elevations. But when a standing wave is flat the water is all moving. This second wave contributes a pattern of velocity.

When the two waves are combined, they together provide the motion and elevation characteristic of a progressive wave.

## Waves that bob and curtsey

Curious things happen when waves of different amplitude travel opposite ways. The bigger wave wins; the wave crests all move in that direction, but they do so in an odd way. They appear to bounce up and down as they travel, being at one moment all low, then high, then low again. They also have the appearance of swooping forward, travelling quickly while they are low and then hesitating while they are high.

I will try it mathematically if you like:

$$\begin{aligned}\text{elevation} &= a\cos(\omega t - kx) + b\cos(\omega t + kx) \\ &= (a+b)\cos\omega t \cos kx + (a-b)\sin\omega t \sin kx \\ &= (a^2 + b^2 + 2ab\cos 2\omega t)^{\frac{1}{2}}\cos(\alpha - kx),\end{aligned}$$

where the angle $\alpha$ in the last expression is not constant but varies by the law:

$$\tan\alpha = \frac{a-b}{a+b}\tan\omega t.$$

The amplitude, which is the square root quantity, fluctuates with time between $(a+b)$ and $(a-b)$. The angle $\alpha$ changes only slowly when the wave is largest, showing that the wave is moving slowly. But when the amplitude is small the angle $\alpha$ changes quickly and the wave hurries along.

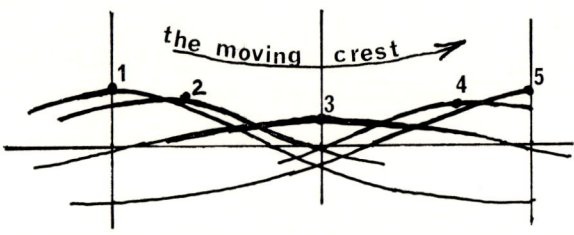

Fig. 4.6. This happens when the reflected wave-train is not so high as the incident wave-train.

I have tried to show in the diagram how the wave sinks and swoops and rises again.

## Short-crested waves are speedy

In the last chapter we found how quickly a wave-train would travel. However, if you add two similar wave-trains travelling in oblique directions you find that you have waves that travel more quickly still.

The diagram is intended to be a view looking down on the sea. I have not drawn the details, but one set of parallel lines is intended to represent the tops of the crests of one wave-train. The other set represents the crests of the other wave-train. I have used a sprinkling of dots to show where the surface is elevated.

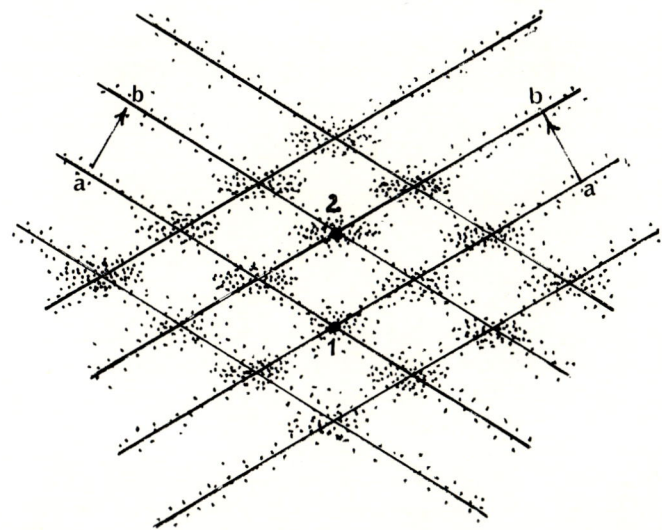

Fig. 4.7. Short-crested waves formed by two wave-trains crossing.

Where any two crest lines cross you will find, of course, that there is a local high patch of water. For instance, the crest lines that I have labelled $a, a$ cross at the point marked 1 and the crowded dots suggest a high patch of water. This happens all over the diagram, wherever crest lines cross. There are hollows too, where troughs cross, so the whole sea would look very wavy. But these waves are short chunky ones. They are sometimes called short-crested waves.

If you wait for one whole wave cycle, the long crests that I marked $a, a$ will have advanced a whole wavelength to the positions $b, b$. Their intersection will then have advanced to position 2. In fact the short-crested wave will have travelled from position 1 to position 2 and this is appreciably more than a wavelength. Short-crested waves are more speedy.

If you are in quiet water near a smooth sea wall you can make waves with a paddle, or by rocking the boat. If you can make these waves travel nearly at right angles to the wall, they combine with their reflections to give short-crested waves that shoot away in the direction *parallel* to the wall. The crests are elongated in that direction and,

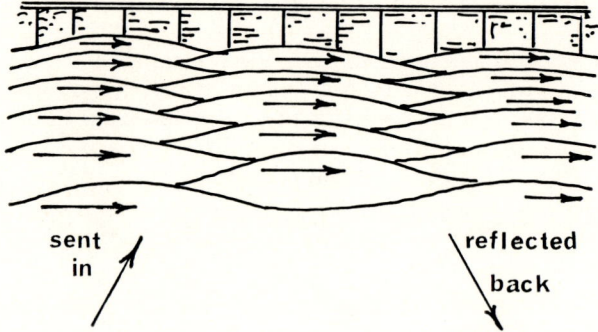

Fig. 4.8. When a wave-train is reflected from a wall almost at right angles to the wave direction, the combination is almost a standing wave, but you can see the crests shoot along very quickly in a direction parallel to the wall.

oddly enough, they travel in that direction, parallel to their length, not at right angles to it.

*Standing waves in a rectangular swimming-bath*

All the water below a crest, or a trough, in a standing wave, moves vertically up and down. You could have a smooth vertical wall there and it would not interfere with the motion.

Elsewhere the water moves horizontally too, but the motion is all in a vertical plane at right angles to the crest lines. So one could have smooth vertical walls at right angles to the crest lines and this would not interfere either.

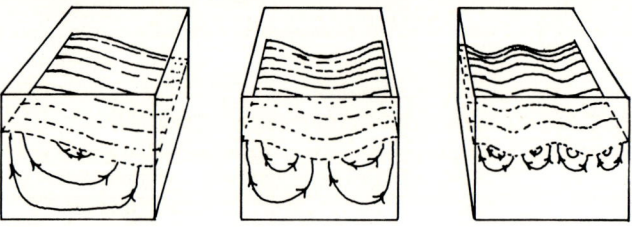

Fig. 4.9. A standing wave trapped in a rectangular basin.

In fact you could have four vertical walls like a swimming-bath and it could hold a standing wave. There would need to be some exact number of half-wavelengths between opposite walls.

I have drawn some pictures of various possibilities, and have included lines showing the direction in which the water moves to and fro as the wave oscillates.

## Standing waves in lakes

Sometimes one can stand on the shore of a lake and see the water slowly rising as if the tide were coming in. But the rise goes on only for a minute or two (it can be several minutes if the lake is large), and then it reverses. The rise and fall will probably be only a few inches. The water level goes on rising and falling in this way for as long as you care to watch. The motion is larger on some days than on others.

If you imagined from this that the water level was rising all over the lake and then falling again you would be puzzled to account for it. But this is not what is happening. As you watch the water level slowly rising at your side of the lake, someone on the opposite shore would see it falling. The water in the lake is slowly swinging from side to side. There is a standing wave in the lake water.

A standing wave in a lake is called a 'seiche' (pronounced, say-sh). It is a Swiss–French word. I suppose that the motion is common in Lake Geneva, though indeed it is common in all lakes where there are winds able to set the water moving.

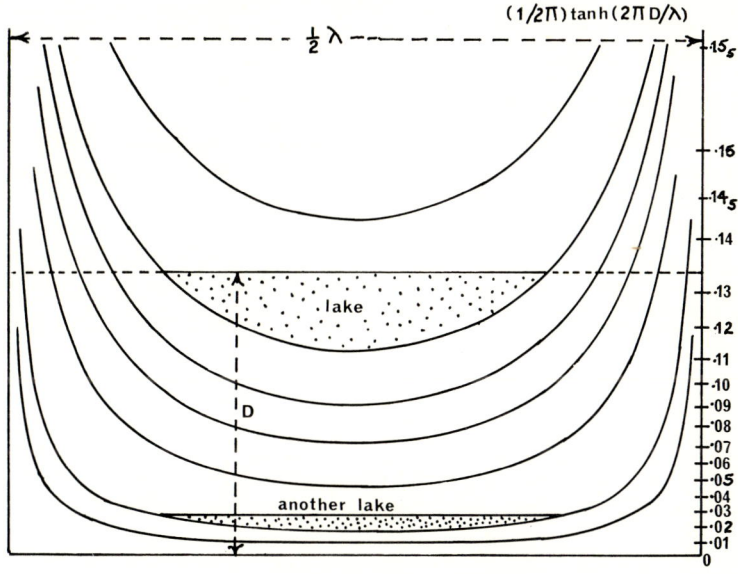

Fig. 4.10. A way of estimating the period of a seiche across a lake or estuary if you know its under-water shape. Pick the shape that fits it best and draw in the water level. Then on this scale the whole width of the diagram is half the effective wavelength and you can calculate the period using the numbers on the right.

Here is one way of predicting the period of a seiche. Figure 4.10 shows the pattern of motion in a standing wave over a flat bed, which is the bottom of the diagram. The diagram covers one-half of a wavelength, the portion between a crest and a trough. When the wave oscillates the water everywhere moves to and fro through quite short distances in directions parallel to the curved ' flow lines '. You could therefore imagine any of these curved flow lines as being the bed of a lake and the water in the lake would oscillate just as if it formed part of the complete standing wave.

So this is what you do ; pick one of the flow lines whose lower part most resembles the underwater profile of the lake you have in mind. Imagine it filled with water, as the diagram suggests, so that you have a picture of the lake section on a reduced scale. You want to know the period of the standing wave. It depends upon the wavelength $\lambda$, which is twice the width of the diagram, and upon the distance between the base (the flat bed) and the level of the water surface in your picture of the lake. I have called this distance $D$. Since you know the scale of your lake picture you can use that scale to find what lengths these distances on the diagram will represent. The formula for the period $T$ of the standing wave is :

$$T^2 = \frac{\lambda}{g} \div \left[\frac{\tanh(2\pi D/\lambda)}{2\pi}\right].$$

To save you the trouble of finding mathematical tables, I have shown on the diagram, for various water levels, the value of the number in square brackets. Put this into the formula together with your scaled-up value for $\lambda$ and the value of $g$ and you have the square of the period of the seiche in the lake.

This argument supposes that the lake is very long in the direction perpendicular to the picture, or else that it lies between parallel vertical walls (which is unlikely). But one has to make approximations in physics, and the estimates this method gives are not far wrong for real lakes.

*Waves in a square basin*

In a square basin, a standing wave going from side to side has the same wavelength and hence the same period as that going from end to end. What will the combination of the two look like ?

I have drawn these two patterns, the standing wave side to side and the one end to end. The middle picture is the sum of the two, on the assumption that their amplitudes are equal. The water is raised up at one corner of the square and has sunk down at the opposite corner. Instead of the nodal line (a line at which there is no vertical motion) being across the width of the basin or across its length, it now runs from corner to opposite corner. If we assumed that the two original standing waves moved in unison, both rising and both falling together, and had equal amplitudes, then the combination would be a simple standing wave in which the water gathers up first at one corner of the basin and then moves across to gather at the opposite corner.

We assumed just now that the two standing waves moved simultaneously, both reaching maximum elevation at the same instant.

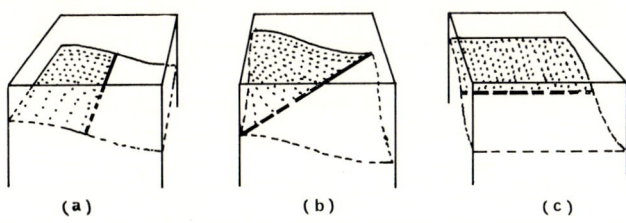

Fig. 4.11. If a basin has equal sides (square) you can have a corner-to-corner wave. The nodal line may also be curved like some of the diagrams in Fig. 4.12.

The usual phrase is that they were 'in phase'. But things don't have to be like that. One could imagine that one wave was at the 'flat' stage while the other was highest. They would then be out of step by one-quarter of a wave cycle, and the usual phrase is 'in quadrature'. The combination then looks quite different. You can readily follow it through the various stages.

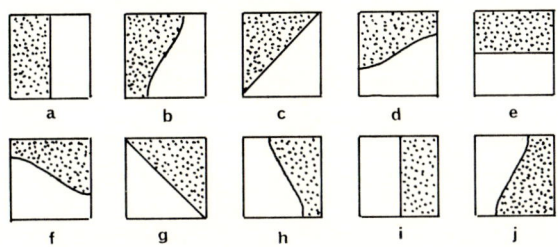

Fig. 4.12. You can even have a wave that rotates within a square basin. Diagrams are at intervals of one sixteenth of a cycle.

We can start with the first wave at its extreme elevation. The other wave is 'flat' so all that we see is the elevation due to the first wave, up on one side of the tank and down on the other.

A moment later, say one-eighth of a cycle later, the first wave will have sunk a little but the second one will have risen. Both waves will contribute elevation, and now we have a diagonal pattern of elevation. But after another one-eighth of a cycle the first wave will be 'flat' and we see only the second; the water is elevated all along one side. You can follow this through with the help of the diagrams. What we have now is a wave moving round and round the square tank. It is a kind of progressive wave except that it goes round the tank instead of moving continually forward in the same direction.

I'm sure you have seen something like this. The water near the middle of the tank doesn't rise or fall at all. It just goes round the centre in a small horizontal circle. But at the middle of the sides, the water goes round in a vertical plane parallel to the wall, and in the corners it doesn't move horizontally at all; it just goes up and down.

The wave can be rather curious if the basin has sides that are nearly equal but not quite. The sides might differ in length by say 10%. If one sets up a corner-to-corner wave, which is easily done by tilting the basin slightly about a diagonal, the first few cycles look quite normal. But because the component waves do not have quite the same period they slowly get out of step and soon one has a wave rotating round the basin. This slowly changes to a corner-to-corner wave but now it involves the other pair of opposite corners. Then the wave begins to rotate again, going the opposite way this time, till it gets back to being the corner-to-corner wave that was set up at first. The process continues.

*Corner-to-corner waves*

Think of two similar standing wave-trains, lying at an angle to one another.

I have sketched the situation as a view from above (a 'plan' view) in fig. 4.13. I have made full lines where the crests will shortly come and broken lines where the troughs will come, but I am thinking of

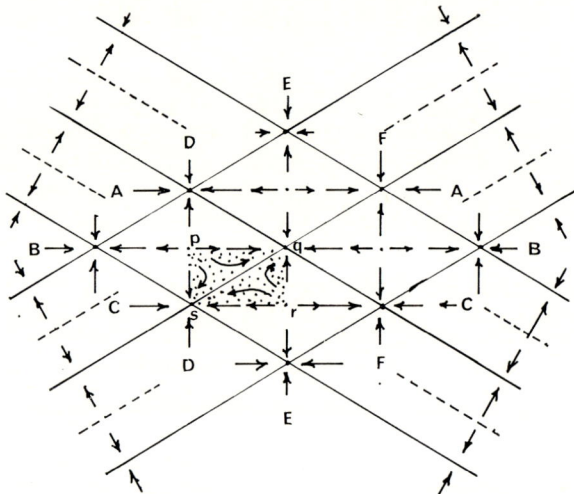

Fig. 4.13. Devising waves that fit a rectangular basin.

the stage before this when the surface is flat but the water is moving. I am assuming that the wave-trains are synchronized; they both are flat at the same moment.

The water is moving, and round the outside of the diagram I have drawn arrows to show the direction of *horizontal* water motion that the two wave-trains separately would produce. The horizontal velocity is small as one gets near to a trough line or crest line and is greatest halfway between them as you see. I am not paying any attention to the vertical component of motion (at right angles to the paper), though there is one of course. You will see in a moment why I need not argue about vertical motion.

The velocity that I have shown, by arrows round the outside of the diagram, extends all across the diagram of course, but in the middle part I have shown what the motion looks like when the two wave-trains are present at the same time, and the velocities add together as vectors. You would be wise to check that I have added up the velocities properly.

This combined velocity pattern looks rather complicated but you will see that there are certain lines along which the water flow is parallel to the lines. For example there is the line AA. I could imagine a thin vertical wall along this line (a wall perpendicular to the paper) and it would not interrupt the water flow, which is parallel to it.

I did not need to worry about the vertical component of water velocity because this would necessarily be parallel to a vertical wall.

There are other lines where I can introduce vertical walls. For example, BB and CC and DD and EE.

To think of the simplest situation I could make four vertical walls to enclose a rectangular pond *pqrs* and the combined standing wave could still continue inside it.

Notice that the water is moving away in all directions from the two opposite corners $p$ and $r$; these will shortly be two isolated wave troughs. At the instant when the water is lowest at $p$ and $r$ it will have risen into wave crests at the other two corners $q$ and $s$. Then of course the water will reverse its flow, go through a flat stage, and then develop crests and troughs the other way round.

This is a corner-to-corner wave in a rectangular tank. Notice the water in the very middle does not move at all. It doesn't even go up and down.

As always, if you wanted to foretell the period of this 'corner-to-corner' wave you would calculate it from the wavelength of the standing waves that make it up. Can you see how to work out this wavelength†, supposing that you know the length and width of the rectangle?

If the walls were placed to enclose a larger rectangle one could have crests and troughs inside the pond as well as at the corners. The

† The wavelength $\lambda$ is given by :
$$4/\lambda^2 = 1/a^2 + 1/b^2,$$
where $a$ and $b$ are the sides of the rectangle.

pond might even be L-shaped or more complicated still. It is necessary only to have the pieces of wall lying on the lines I have pointed out, where the water moves parallel to the wall.

## Waves in a circular basin

I used to work in a building where machines were running. A properly balanced machine should not vibrate, but no machine is perfect. These gave a slight tremor to the table at which I sometimes drank a cup of tea. Very handsome and quite tiny patterns of concentric rings often formed on the surface of the tea in the cup. They were standing waves, but in this case the crest lines were circles, not straight lines.

An engineer often has to deal with circular patterns of waves. To describe them mathematically he uses ' Bessel functions '. I want to show that the circular waves can be built up from our straight long-crested waves, so there is nothing strange about them even though they do illustrate one kind of Bessel function.

I also want to find a formula for the speed of travel of very short waves (ripples). The formula we found before doesn't work, because in such small-scale waves the surface tension has a big effect.

## Making up a circular wave pattern

Start with a standing wave that has long crests and troughs lying, say, north and south. Now add another similar standing wave where crests lie east and west. The result is pictured in the first diagram.

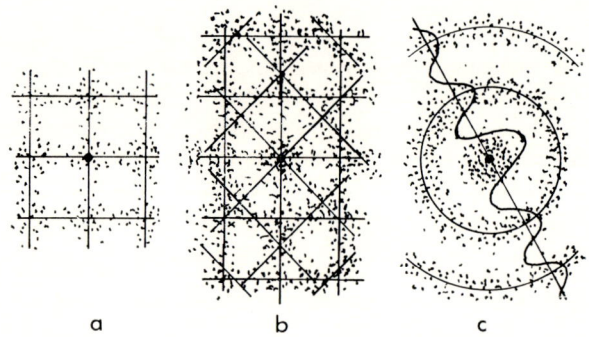

Fig. 4.14. Making up circular standing waves by superposing straight ones ; (*a*) shows part of the pattern made by two standing waves at right angles ; (*b*) shows part of the pattern made by four standing waves at angles of 45° ; (*c*) shows the circular standing wave and also the shape of its profile along any diameter.

Now add two other sets of standing waves that cross the others diagonally. The result is pictured in the second diagram. I am assuming, as you can see, that all these sets of waves have crest lines passing through some point that forms a centre. I am also assuming that they all rise and fall together.

If you go on in this way adding further sets of waves at intermediate angles you can see that the result would become more and more like the perfectly circular pattern shown in the third diagram.

In this circular pattern the greatest elevation comes at the centre, for all the straight waves have crests there. The surrounding rings of crests grow gradually less high as one looks farther from the centre.

This circular pattern is a standing wave. The crests duck down to become troughs and then reverse into crests again. There are moments in between when the wave is flat, and then the water is all moving.

*Fitting the circular wave to the cup*

At the vertical sides of the cup, the water must move vertically (or to be more precise it must not move radially). Vertical motion in our circular pattern occurs exactly at a crest line or a trough line. I will not try to prove this but you will agree that it seems reasonable.

So the circular wave patterns that fit the cup must be of such a scale that a crest or a trough comes just at the sides.

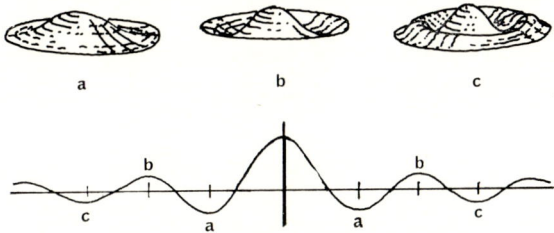

Fig. 4.15. Fitting the circular standing wave into a tea-cup. The curve shows the profile through the centre. One scales it to bring a crest or a trough to the rim.

I can make up circular patterns that are wide or narrow merely by choosing an appropriate wavelength for the straight waves. So, in the diagram I have drawn some profiles of different circular standing waves that would fit a cup of a given diameter.

If you wanted to describe these profiles mathematically you could write :

$$\text{elevation at radius } r = J_0(kr),$$

and an engineer would know what you meant. He calls $J_0$ a 'Bessel function of the first kind of zero order' and he has numerical tables showing how it varies with $kr$, which he calls the 'argument'. The profiles I have drawn are just graphs of this $J_0$ function, and I used the tables to draw them. I had to adjust the scale to make a maximum or a minimum (a crest or a trough) come just at the outer edge.

The circular standing wave bobs up and down, of course, so it would make a better description if I said so by including a factor like $\cos \omega t$:

elevation at radius $r$ and at time $t = J_0(kr) \cos \omega t$.

I ought also to include an amplitude factor. The mathematical tables are arranged so that $J_0$, which is our profile, rises just to the value 1 at the middle. The cosine, too, has an amplitude of 1. So if I wished to say that the wave at the very centre had an amplitude of half a millimetre (or whatever it might be) I should multiply by this amplitude. But I won't trouble to write the formula again.

Finally, you will agree that different circular patterns bob up and down at different rates. They must have the same frequency as the straight standing waves of which we built them up. How can we tell what this will be?

The mathematics is arranged so that the $k$ in this formula is just the $k$ (the wave-number) of the straight waves one could use to make up the pattern. This should tell us their frequency.

Then how do we find $k$? Well, it comes from the scaling I had to do. For instance, in drawing the first profile I wanted the outside edge to be the first trough of the circular pattern. When I looked at the mathematical tables I found that the first minimum of $J_0$ came where the argument was equal to 3·83. But I know that the argument is just $kr$. So when I know the radius of the cup I know the wave-number of the straight waves. It is:

$$k = 3\cdot 83/r.$$

Then knowing the wave-number $k$ I should be able to deduce the frequency.

*Other circular waves*

There are other types of circular waves. I have made sketches to show three of them. These can all be thought of as being composed of straight waves.

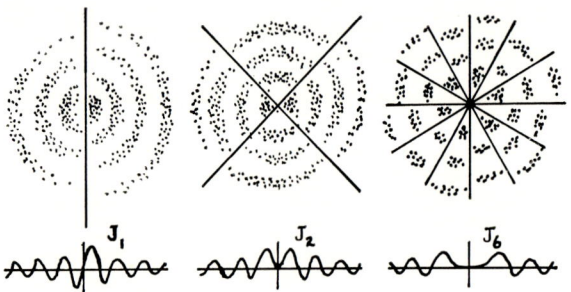

Fig. 4.16. Some other types of circular waves and their profiles. The straight lines mark places where there is no rise and fall.

Perhaps it is easiest to start with progressive wave-trains, travelling in a variety of directions. In the last section I sent in equal waves from all directions. This time I will use waves with different amplitudes.

I will imagine a long-crested wave travelling across towards the east. This will be at full strength. As I look round more towards north I see other similar wave-trains travelling across but they are weaker. In fact I am imagining that the amplitude of each wave-train is decided by the cosine of the angle between its direction and east.

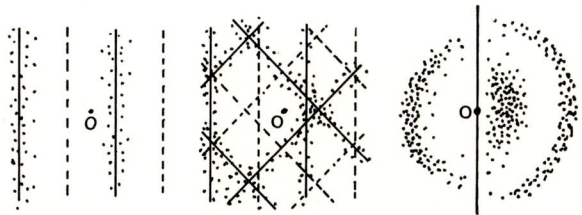

Fig. 4.17. Making up the $J_1$ standing wave.

When I look north there is no wave at all going that way because $\cos 90°$ is zero. But as I look farther, westerly this time, the cosine tells me that the waves will have a negative amplitude. I mean by this that at the moment when the waves with positive amplitude have crests at the centre point, these waves travelling west have troughs there, a negative elevation. And so it continues. When I look west there is a strong (and negative) wave but looking more south the wave is weaker. Nothing travels due south, but as I look more east the wave builds up again and is positive now.

I think you will agree that because of the reversal in sign, the combination of all these waves gives exactly zero, always, at the centre point. But they don't cancel everywhere. For instance, the east and west-going waves are sure to produce a standing wave that I have illustrated in the first sketch. They cancel at the centre point but this merely means that there is a crest to the east of the centre and a trough to the west as the diagram shows. The story is the same for other sets of opposite directions. In the second sketch I have included the two directions at 45° to east. When all the waves are added in, the composite picture shows a circular pattern. It is the same type as the first one in the previous diagram, but having taken a larger wavelength I cannot show so many crests.

This is a standing wave; the crests duck down and become troughs while the troughs rise up into crests, and this reversal goes on continually.

A feature of this pattern is that the profile along every diameter has the same shape but the waves are weaker along sections taken more nearly north–south. I am not able to prove it to you, but the wave amplitude in the various sections just depends on cos $\theta$.

I sketched the profile along a diameter in fig. 4.16. Notice that it is zero at the centre. An engineer would call this profile a 'Bessel function of the first kind and first order', or more briefly, just $J_1$. He has mathematical tables of this curve too. If you wanted to describe the whole circular pattern, bringing in the variation with angle, you would write:

$$\text{elevation at radius } r, \text{ angle } \theta = J_1(kr)\cos\theta.$$

You might wish to put in a factor like $\cos\omega t$ to describe its fluctuations in time, and you would need some numerical coefficient to make the waves the right height.

Fitting the wave to the cup and finding its frequency goes on just as before.

I won't go on to build up the other patterns of fig. 4.16, for you can see how it would be done. In the second pattern the factor is $\cos 2\theta$ and in the third I chose $\cos 6\theta$. All these more complicated patterns are zero at the centre point. Indeed, in the last pattern there is quite an area around the centre in which the waves are very weak. It's as if the sectors were so narrow that the waves could not penetrate to the ends.

You may wonder about 'Bessel functions of the second kind'. Are there such things and are there waves like them? Yes, there are. Indeed, they look very like the ones we have seen, circular crests, either quite symmetrical or arranged in sectors. But you need not think about them unless there happens to be some central obstruction with waves round it. I would never see them in my tea-cup.

Notice, too, that if this 'second kind' need a central obstruction I cannot make them by adding up a variety of straight waves. The central obstruction would make complicated reflections. So I won't try.

## The speed of ripples

I found a speed for deep-water waves in the last chapter. It was:

$$c = \sqrt{(g/k)}.$$

Waves in a cup are so short, however, that they do not obey this formula. I thought that I might explain why.

The surface of water behaves as if it were a thin elastic skin with a tension, the 'surface tension'. In very short waves the surface becomes sharply curved. Such bulges and hollows in an elastic skin require a difference between the pressures on the two sides, which are the air and the water. The pressure in the air may be the same everywhere but now the pressure in the water at the surface will vary.

I don't propose to work this out in detail, for I said in the last chapter that we had come to the end of the mathematics. You can readily see, however, that the effect of this skin will be to make the waves travel faster. Look at it like this. Gravity alone, the hydrostatic effect, would mean a pressure that was lower at the level of the crests than at the level of the troughs. To counteract this we had to make the waves travel. This somehow produced an extra positive pressure at the crests and a negative one at the troughs and we adjusted the speed until the two effects together provided a pressure that was the same at both places. Now that we want to take surface tension into account we need to have still more positive pressure on the crests in order to account for the upward bulge in the skin, and of course a reverse effect in the troughs where the skin is drawn down. We must therefore make the waves move still faster.

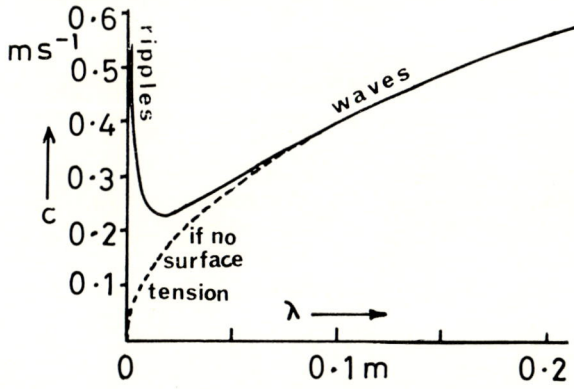

Fig. 4.18. The calculated curve of wave speed against wavelength when one remembers that water has 'surface tension'.

The formula turns out to be:
$$c = \sqrt{(g/k + Tk/\rho)},$$
where $T$ is the value of surface tension.

If waves are very short, their wave-number $k$ is large, and it can happen that the second term is more important than the first. It is surface tension rather than gravity that makes the waves move. We usually call these short waves 'ripples'.

You will notice that in the second term the speed increases with $k$. The shorter ripples are, the faster they move. This is quite the opposite of what happens with ordinary sea waves; there the shorter ones travel more slowly.

*The frequency*

We have been discussing formulae for the speed $c$. If you are interested in frequency, remember that for any sort of wave it is true that
$$\omega = ck,$$
so you can easily write formulae for $\omega$ or for the frequency, which is $\omega/2\pi$.

# CHAPTER 5
## groups of waves

PEOPLE who build ships work to a design, and people who design ships need to know, among other things, the best shape for a ship to be.

To help to decide this, there are engineering laboratories that make ships in model size, 2, 3 or even 4 metre long and tow them through water to find what force is needed. I should explain that the models are usually made of paraffin wax which is cheap to buy, will float as a solid lump and is easy to pare down to just the right size and shape.

In these laboratories there are long tanks of water, perhaps 5 metre wide, 3 metre deep and 200 metre long. An overhead carriage can run on rails along the whole length of the tank, towing the model ship and carrying instruments that record everything that may be interesting about the way in which the model behaves.

Since ships at sea can expect to meet waves, these long tanks can make waves too, and that is why I am talking about them. At one end is a wave-making machine, perhaps a large paddle of wood and iron that can be regularly moved to and fro in the water. At the other end the tank has a sloping 'beach' that breaks the waves and prevents reflections. Then of course you can tow the model ship against the waves or in the same direction and watch for any behaviour that you would dislike if it were a real ship and you were on board it.

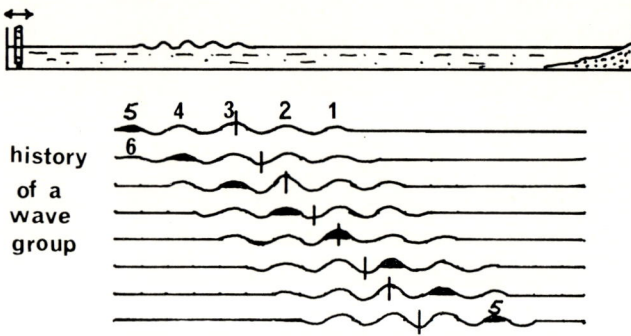

Fig. 5.1. A group of waves travelling down a tank. One particular wave crest, number 5, is shown in black in successive profiles.

But given such a tank and a wave-making machine one can make experiments just on waves to see how they behave, and I want to tell you now about one of these.

With the tank vacant, all models removed, you drive the wave-maker to and fro about eight times and you find that you have started a group of waves.  There are only four when you look at them, which seems surprising, but let's carry on.  You see these waves advancing down the tank ; every time you look at them there are four, or perhaps sometimes five if you count a very low one at the end.

Yet if you stand at one place beside the tank and count all the waves as they pass you, you find that eight waves pass you.  Anyone who does this agrees that eight pass.  The waves pass all the way down the tank and break on the beach ; eight waves break on the beach.  This seems to link with the fact that the wave-making wedge cycled eight times, yet it seems odd that if you had photographed the waves in the tank the photograph would have shown only four.

If you had made a ciné film of the waves, perhaps from the moving carriage so that you could keep them in view all the time, you would have seen the reason for the curious disagreement in the counts.

As the cluster of four waves moves forward a new one begins to grow up in the rear and at the same time the leading wave grows less as it travels, and fades out.  Then another begins to form again behind the last, which has now grown to full size, while again the leading wave grows less and fades to nothing.  This goes on continually.  The result is that the region of water that has waves on it does not move down the tank so quickly as do the individual waves.

*The motion of waves in a group*

I think that this curious behaviour of water waves may be new to you.  It does not happen with sound waves, nor with waves travelling along a string or along a spring.  One does not notice it even when watching breaking waves from the shore, so I have made a diagram to illustrate the behaviour, and have put numbers on each wave crest to show how they move.  As wave number 5 builds up, wave number 1 fades away, as 6 builds up, wave 2 fades away, and so on.  Then if you count the waves passing a certain place there are eight or nine, though four or five is the number you can see at any one moment.  This is what happens when waves are in deep water (deeper than half the wavelength, say).

You may think it odd that waves can disappear or appear from nothing in this way.  Remember, however, that a wave is not a material thing, like water ; it is a pattern of behaviour.  Nevertheless, you say, a wave has energy and this is 'conserved' ; the energy of the wave cannot fade out in the way the waves seem to do.  This is right.  When a sound wave travels through air, each piece of air does work on the air in front of it ; it transfers energy forward and one can calculate how much is transferred per second or per wave cycle.  It turns out to be just the right amount to supply the waves that have passed forward.  A similar calculation (which I am not going to make here) shows that in waves on deep water the energy is not fed forward quickly enough to supply the leading wave. So the leading wave dies down while a new wave grows in the rear from the energy left behind.

## A formula for group velocity

Instead of thinking about energy we can use a different argument. Think of a regular wave-train. I have pictured its profile as a broken line in the diagram. There is no question of waves disappearing here; the wave-train by itself would advance without changes in amplitude†. Now think of a second wave-train added to the first

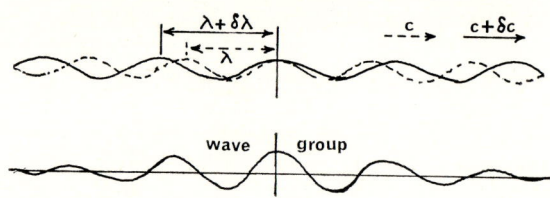

Fig. 5.2. An argument to find how quickly a wave group travels.

one, the wavelength of this second train being slightly greater than the wavelength of the first, say $\lambda + \delta\lambda$ instead of $\lambda$. I have shown it as a full line. We can suppose for the purpose of the argument that two of the crests in these wave-trains happen to agree precisely in position. When the two wave-trains are added it is here that one will find the largest wave of all because elsewhere the wave-trains are somewhat out of step, owing to the wavelengths being unequal. For example, if we look at the two wave crests immediately following, one lags by a distance $\lambda$, the other by a distance $\lambda + \delta\lambda$, so that here the crest of the longer train is a distance $\delta\lambda$ behind the crest of the other.

But the train of longer wavelength travels at a different speed, say $c + \delta c$ instead of $c$. In a short time this lag $\delta\lambda$ will have been made up. The time required for this to happen is of course $\delta t$, where

$$\delta t = \delta\lambda/\delta c.$$

Both wave-trains will have moved forward during this time, of course, but notice that the place where the crests agree has dropped back relative to them by one whole wavelength.

So the highest wave is now a different wave, and because we look on the highest wave as marking the middle of the wave group, this group

† You can think, if you wish, that as each wave moves forward, half the energy it needs is fed forward into the new water from the water behind while half the energy of the preceding wave remains behind and makes up the required total.

has advanced a shorter distance than the waves themselves. Indeed, the distance is:
$$\delta x = c\delta t - \lambda.$$
The first term is the distance gone by the wave crests[†] in the interval $\delta t$. The second term is the one wavelength the group has dropped back. So the speed of the packet or group is:
$$\delta x/\delta t = c - \lambda/\delta t$$
$$= c - \lambda(\delta c/\delta \lambda).$$
The speed at which the group advances is known as the 'group velocity'. We can call it $U$ and since $\delta c$ and $\delta \lambda$ can be quite small we can write:
$$U = c - \lambda(\mathrm{d}c/\mathrm{d}\lambda).$$
This is the 'group velocity', the speed at which a wave group travels.

## *In deep water*

Let's see how this formula works for waves in deep water. We found that the wave speed was given by:
$$c^2 = g\lambda/2\pi.$$
Rearrange this to get:
$$\lambda\,\mathrm{d}c/\mathrm{d}\lambda = g\lambda/4\pi c = \tfrac{1}{2}c.$$
Then $U$ turns out to be just $\tfrac{1}{2}c$. The groups travel at only half the speed of the individual waves. This is what I took to be true when I described the group of waves moving in the long tank.

## *In very shallow water*

We found a formula:
$$c = \sqrt{(gD)} \quad (D \text{ the depth of water}).$$
But now the wave speed does not depend at all on the wavelength. On looking at the formula for group velocity you see the $U$ and $c$ are equal.

This is why you don't see waves appearing and disappearing when you look at the sea from the beach. The water is so shallow that the waves and the wave groups travel forward at the same speed.

## *A graphical method*

If you draw a graph of wavelength against wave speed you can use it to read off the group velocity too.

---

[†] Strictly speaking, $c$ is the speed of the slower wave-train, and the other has a speed $c + \delta c$, but I am assuming that the difference in speed is quite small.

Choose a point on the graph at the wavelength you are interested in and draw a tangent there. Extend the tangent line so far that it cuts the velocity axis, and the value you read there is the group velocity for that kind of wave.

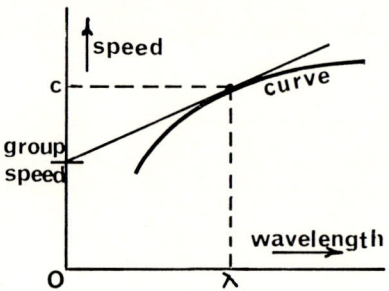

Fig. 5.3. Reading group speed from a curve relating wave speed and wavelength.

I think that you will see why this method works if you look at the formula :
$$U = c - \lambda dc/d\lambda.$$

## Heavy surf at Barbados

If you look for the small island of Barbados (about 10 miles by 20) on a chart of the North Atlantic Ocean, you will find it some 250 miles north of the delta of the Orinoco. It is the most easterly of the large group of islands that enclose the Caribbean Sea. Fishermen in Barbados use small boats which they leave at anchor inside the submerged coral reef that fringes the shore. Waves coming from the sea usually break as they cross this reef, so boats at anchor, and houses near the beach, are protected unless the waves are violent.

Violent waves can develop at Barbados quite unexpectedly, at times when the weather is fine and calm. We now know that such waves can come from storms more than 1000 miles away to the north. The situation at Barbados is typical of islands in big oceans, but I have spoken of Barbados because there is now clear evidence in the matter. During the International Geophysical Year (the ' I.G.Y.') which was held in 1957–58, an instrument for measuring the rise and fall of water in waves was put in the sea off the north-eastern coast of the island, outside the coral reef. It was connected to the shore by armoured electrical cable and it provided continuous records from which one could read values for the average wave height and for the typical period of the waves. Period is important because one can infer from it the speed at which the waves will have travelled from the place where they originated.

High waves came unexpectedly on two occasions while the wave recorder was working and I will tell about one of them. This was in October 1958. The sea had been almost calm on the 24th. During the night the waves grew greater and by midday on the 25th the fishermen said that this was one of the worst occasions they remembered. The average wave height at the measuring instrument was 2 metres but this was in fairly deep water, and breakers on the reef were occasionally 4 metres high.

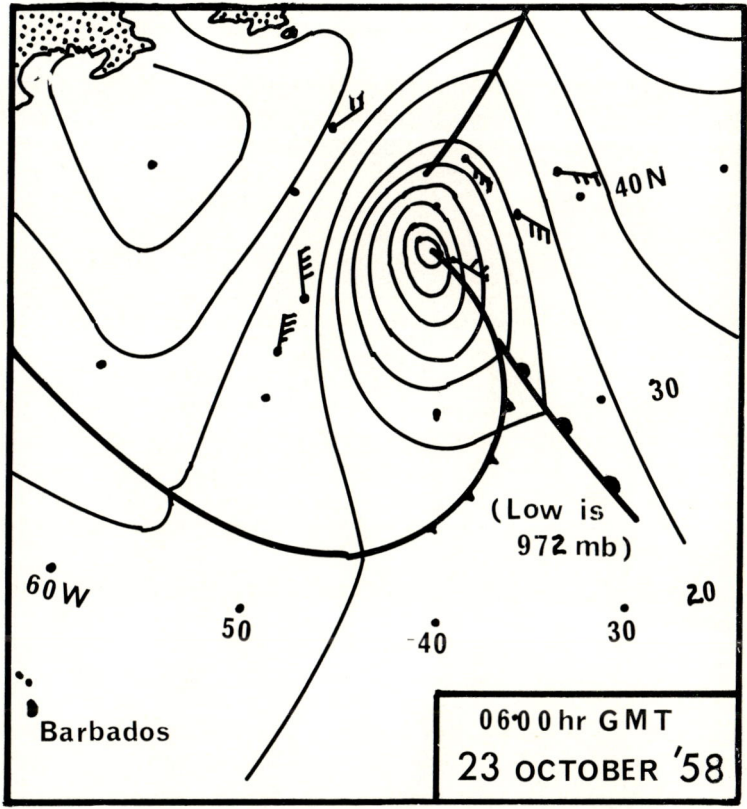

Fig. 5.4. The storm that sent big waves to Barbados.

Then matters grew much worse. Next day at 05.00 hrs GMT the average recorded height was 5 metre and waves occasionally 10 metre high were passing the reef. The period of the waves at this time was 17·5 second. Fishing boats of all sizes were thrown up on the beach and many houses were flooded with water and sand. The violence of the waves very gradually grew less and after two more days they ceased to be dangerous. The weather remained fine and calm throughout.

Weather maps are continually made of the North Atlantic, for the guidance of ships and aircraft, and are based on radio reports made by them every 6 hours. When the scientists on Barbados looked at these maps (later on of course when all was over) they saw that an intense storm had developed quite quickly in about 12 hours in the middle of the ocean at the latitude of New York. It remained for about 36 hours, not changing its position much, and then died away. I have made a sketch of the weather map at the stage when the storm seemed most intense. The thin lines show the contours of equal air pressure ('isobars'). The map is based on weather observations made only at those places where there are arrows showing the direction of the wind at that time and its speed (one feather for every 10 knots).

Could this storm have generated the Barbados waves? One must look for winds towards Barbados with a speed at least equal to the speed of waves of period 17·5 second. According to our deep-water formula this is 27·3 metre per second.

I will write this speed in knots (sea miles per hour). It is convenient to reckon in knots when using charts, because 1 sea mile is equal in distance to 1 minute of latitude, 60 to a degree, and degrees are marked on charts. The wave speed is 54 knots.

The highest reported wind speed at this time was 40 knots but 45 knots was reported 12 hours before and 50 knots a day later. It is also possible to estimate the wind speed from the closeness of spacing of the isobars. The winds blow approximately parallel to them†. It appears possible that waves of period 17·5 second travelling towards Barbados could have been generated within the region a little north-west of the storm centre.

The distance on the chart from the middle of this region to Barbados is equal to 31 degrees of latitude, or 1860 sea miles. If the waves started at 06.00 on the 23rd and arrived at 05.00 on the 26th (an interval of 71 hours) then their average speed was 26 knots.

This fits very well. We know that though individual wave crests would travel at 54 knots, the region of sea disturbed by the waves would advance at only 27 knots, the group velocity, which is half the phase velocity for deep-water waves. It seems clear that the waves came from this storm.

Working from the weather maps in this way it is now possible to give warning to people on the Caribbean islands of times when heavy surf may be expected from North Atlantic storms.

† You may think it odd that the winds blow along the contours of equal pressure rather than across them, from high towards low pressure. Indeed the air is moving from high to low-pressure regions but, like the water in the round basin, it acquires an increased rotation as it converges, to such an extent that it spirals inward in a direction almost parallel to the contours.

## Swell from distant storms

The earliest detailed study of long-distance swell† was made in Britain about 1945. One problem was to make an instrument that could be put in the sea without being swept away in the first storm. The solution was to put it on a tripod standing on the sea-bed where the water was 15 metres deep. The instrument was a metal box, actually a bell-shaped metal casting, with one side made of thin metal sheet. This moved slightly inward and outward as the pressure of the water varied when waves passed over, and the deflection was

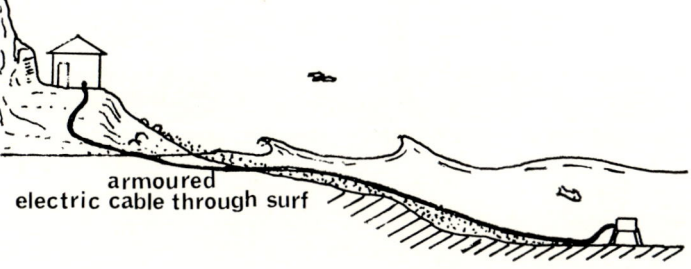

Fig. 5.5. One way of making automatic records of the arrival of waves at a coast.

detected electrically through a cable connecting the instrument to the shore. Continuous records were made in this way of the fluctuating pressure. The instrument was sited near Land's End so that it would be exposed to swell coming from a wide area of the North Atlantic Ocean.

The second problem was that of distinguishing the different frequencies that might all be present at one time in the wave record, for it is clear that the waves arriving at any one time could have started at many different places and in different storms. So the instrument near Land's End was made to record the fluctuating pressure, not as a fluctuating line but as a fluctuating black profile on white photographic paper. Each record in turn (that is a run of 20 minutes taken every 2 hours) was placed round the broad rim of a wheel mounted on ball bearings. The idea was to rotate the wheel and get the history of pressure repeated at high speed again and again as the wheel revolved. A narrow line of light was focused across the rim of the wheel at one place, so that, as the wheel revolved, the varying width of the white part of the profile produced a varying intensity of reflected light. A photocell was used to give an electrical signal varying in imitation of the varying reflected light.

† Waves that have travelled away from the storm region where they arose are called 'swell', presumably because of their smoother appearance. I should have used the word 'swell' for the Barbados waves.

Finally, the electrical signal from the photocell was made to drive a 'vibration galvanometer', one that had a natural tendency to vibrate at 100 Hz but would make very little response to other frequencies. Now you can see the idea, the wheel turns, the photocell reproduces the pressure history at high speed again and again, and if any of the

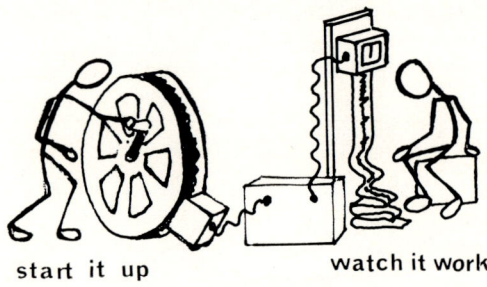

Fig. 5.6. One way of picking out the components of different frequency in a record of waves.

recorded waves happen to have a period that finally leads to a frequency of 100 Hz, then the galvanometer vibrates with an amplitude proportional to the amplitude of pressure in the wave.

The labour-saving aspect of the system was that the wheel was turned to a high speed, perhaps three revolutions per second and then left. Under the slight friction of the bearings the speed slowly and smoothly decreased and so without further effort on the part of the experimenter all the various wave frequencies present in the record came one by one into tune with the galvanometer and activated it. The behaviour of the galvanometer was being recorded automatically. You might say that the galvanometer drew a picture of the frequency spectrum of the waves, in the sense that it looked in turn at all the types of wave in succession and recorded their amplitude.

Figure 5.7 shows a sequence of 'wave spectra' found on one occasion. They cover an interval of 3 days. They show arriving at Cornwall, swell that had started in a storm off Cape Horn, between 6000 and 7000 miles away. The first waves to arrive are those with a long period, 22 second. This is because they travel more quickly than the others. As one goes on from record to record the arriving swell shows a shorter period and at last is lost as it becomes obscured by other short-period swell arriving from places in the North Atlantic.

You may feel surprised that waves from a storm off Cape Horn can reach Cornwall. I assure you that the people who made the experiments were surprised too. They were also suspicious, for they did not wish to be like the Queen in *Alice through the Looking Glass* who prided herself on her ability to believe the most improbable things. I had better explain the arguments that were used.

First it had been seen that swell from storms in the North Atlantic followed a similar pattern; the first swell to arrive had a long period and then the period slowly decreased as time passed. Indeed, it had proved possible to look at the rate at which the period of the swell was

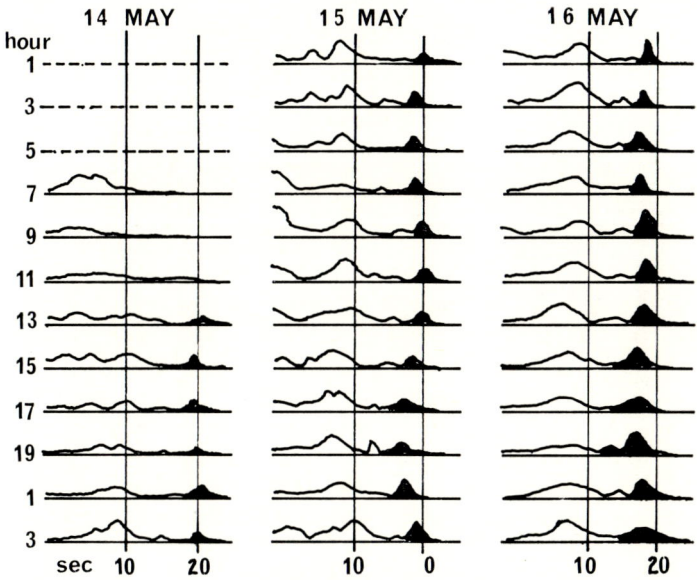

Fig. 5.7. A sequence of wave 'spectra'. The hump that marks the long-period waves arriving from the storm near Cape Horn has been inked in to make it more obvious. This sequence shows only the first 3 days.

changing and to say that it must have come from a storm happening at such and such a time at such and such a distance from Cornwall. Then one looked on the weather maps of that time and there indeed was the storm, at the right distance away. It was very like the way in which one can judge the distance of a thunderstorm by counting seconds between seeing the lightning flash and hearing the thunder; every 5 seconds means about a mile in distance. With swell one counted in hours and reckoned distances in hundreds of miles. The system worked very well. Now, in this series of records the period of the swell decreased very slowly. showing that the storm must have been at a great distance. The estimated distance was between 6000 and 7000 sea miles.

Second, one naturally asks whether there is an open path between Cape Horn and Cornwall. One expects that waves on a round Earth will travel along 'great circle' paths. If you take a terrestrial globe you can stretch a string between Land's End and the South Shetland Islands and it lies on open sea all the way. The distance is 120° of latitude, or 7200 sea miles. One might very well have a storm a little nearer, say at the latitude of Cape Horn or South Georgia. Notice that the month is May, the start of winter in the southern hemisphere.

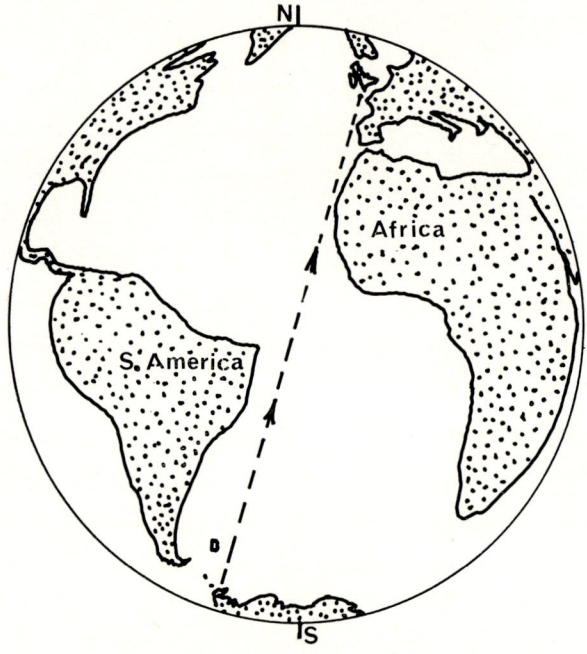

Fig. 5.8. The 'great circle' path of the waves from Cape Horn to Britain.

Third, if you look at the records again you will see that the period of the swell does not decrease in a regular way. The period oscillates. It even increases at times, which is a very odd behaviour. Unless one can give a reasonable explanation for this, the whole story remains unconvincing.

There is a reasonable explanation and it depends on something that happens when the swell is quite near to Cornwall. The sea around Britain is quite shallow, 200 metre or less. Indeed, every continent is surrounded by a border of shallow sea. It is as if the land had been eaten away by the action of the waves, leaving a flat 'continental shelf' extending out under water to a distance of perhaps 5 or perhaps 50 miles from the visible beach. Then at the edge of this shelf the sea-bed goes down quite steeply (a slope of 1 in 10, which would seem quite steep if you had to climb up it for 20 miles), down to a depth of about 2 miles (4000 metre) where it flattens out as the bed of the deep sea. I might say here that this sloping edge of the continental shelf is carved into gullies and canyons by underwater avalanches of rock and mud going down to the deep sea.

But my point is that swell travelling to Cornwall from the South Atlantic would cross 150 miles of shallow continental shelf and would take 5 or 6 hours to do so. The flow of water in tides is of course much increased where the water is shallow and in this region the tidal currents reach about 1 knot, setting north-east and south-west which is along the direction of travel of the approaching swell. These tidal streams could change the period of the swell.

The idea that the wave period can be changed strikes many people (including oceanographers) as being a wrong one, so I will try to explain how I think it can happen.

First of all, a case where the wave period is not altered. Think of a regular succession of waves coming in from deep water, and they enter a region of sea where there is a tidal stream flowing in the same direction. I have pictured this in the upper diagram of fig. 5.9.

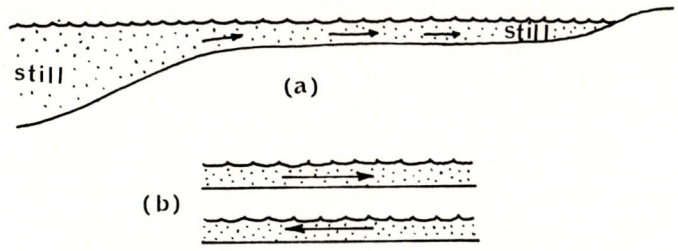

Fig. 5.9. Why the tidal streams made the wave period fluctuate slowly.

After the waves have gone along with the stream some distance they reach the shore.

The waves keep coming and the water keeps flowing. You may be wondering where I think the water goes to, but you can imagine it changing direction and flowing parallel to the shore when it gets near to it.

This picture is a 'steady state' in the sense that things remain the same, and in that case I think that the period of the waves will be the same whether one measures it in deep water or on the shore. The argument is that this is a regular wave-train where all the crests keep their identity, and cannot fade out or new ones come into existence (as happens when there are mixtures of wave-trains). Consequently, if fewer waves per minute arrived at the shore than came from the open sea, it would imply that they were accumulating on the intervening water, becoming more crowded there. Or if more struck the shore than were supplied from deep water, then the waves would need to be opening out and becoming fewer in total number on the intervening water. This would not suit the idea of a 'steady state', so

I think that the wave period would be the same at sea as at the shore.

You may point out that the waves on the stream would be travelling more quickly (relative to the sea-bed) than the waves on the deep water, because the stream helps to carry the waves along. I think this would be true, but it need not mean that more waves per minute pass any fixed point. I think that the waves on the stream would be spaced farther from crest to crest. It would be like a stream of cars along a road. Where the road was more difficult so that the cars travelled more slowly, they would be crowded together. Where the road was easier so that the cars travelled more quickly they would be farther apart, and at whatever place you watched, you would see the same number of cars passing you per minute.

Then how do I imagine the period of the waves can be changed?

The tidal stream is not really constant. It varies with the tides, and 6 hours later would be flowing the opposite way. I am thinking of a case in which the region of tidal stream is so wide that it takes the waves several hours to cross it, perhaps as much as 6 hours. Then they would enter the tidal stream when it was flowing one way and emerge when it was flowing the other way.

The process of entering the tidal stream would not alter the period relative to a fixed point; we agreed on that. The process of emerging from the stream into quiet water near the shore would not affect the period either. But what about the time between, when they were riding on the stream, and the stream gradually reversed its direction?

Imagine the waves entering the stream at a time when it is flowing in the direction in which they are going. The wave crests become more widely separated. Then, as they travel along a stream whose speed is the same everywhere, the stream slowly comes to rest and begins to flow the other way. Because all the crests must have the same speed at the same moment their separation ($\lambda$) could not change and their speed through the water would not change. But the change in the direction of the stream would change their speed relative to the ground since the water drift now reduces their speed relative to ground instead of adding to it. The same wavelength and a reduced forward speed mean fewer waves per minute passing a fixed point. So the period has increased and the increase would be noticed when the waves came ashore.

This increase does not persist indefinitely. Soon the waves that enter the stream find the stream opposing them. They travel more slowly and are more crowded in consequence. These waves, however, see the stream slowly stop and begin to flow in the direction in which they are travelling. They remain crowded but they are carried forward. More waves per minute pass any fixed point. Their period is reduced and this persists when they leave the stream and arrive at the shore.

I think therefore that a very wide region of tidal stream could make the wave period vary with the cycle of the tides, the period being sometimes a little greater than the wave period in deep water and sometimes a little less. The fluctuation would have a cycle of 12 hours, or more exactly, the time cycle of the tides.

If you look at the records in fig. 5.7 you will see that the fluctuations do indeed have a cycle of about 12 hours and this agrees nicely with the idea that it is brought about by the tidal streams.

It does indeed seem that this swell started near Cape Horn.

*Finding wave direction too*

The most recent experiments on far-travelled swell were made in 1959 by a group of people working near San Diego on the coast of California. The site where they made their wave records was exposed to swell originating almost anywhere in the Pacific Ocean.

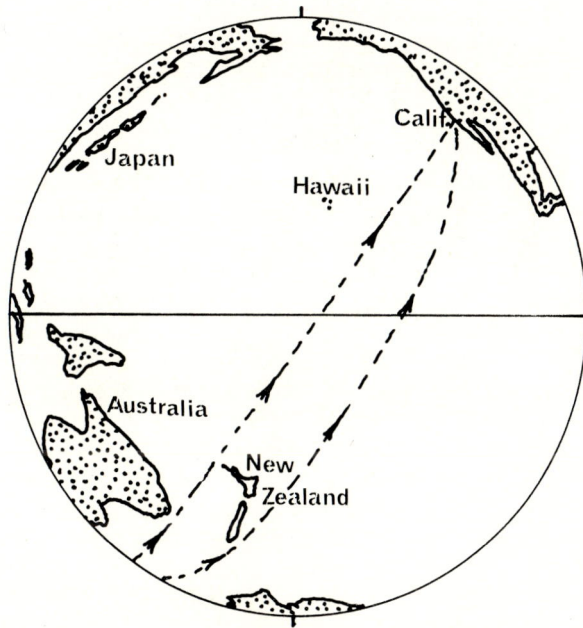

Fig. 5.10. The passage of waves to California from storms south of Australia. This distance is nearly halfway round the world.

A new feature of these experiments was that three wave-detecting instruments were set down at the corners of a triangle. With this arrangement you can see that it became possible to distinguish the direction from which any swell was coming by noting the sequence in which each wave arrived at the three instruments. This idea worked very well. In most cases it was possible to judge the direction of travel of the swell to within 5°.

The instruments themselves were more sensitive and more exact than those used in the early work that I have already described. The method of picking out the components of different frequency in the swell was also greatly improved; the pressures were recorded every 4 seconds by punching the number on paper tape using the binary† system, and the information on the punched tapes was later read by a digital computer which then calculated a frequency spectrum. It also compared the pressures at the three instruments so as to give an estimate of the direction of travel of the swell.

Swell from many different storms was identified during the 3 months in which the experiment continued. In three cases it appeared that the swell must have travelled nearly halfway round the world, starting east of Australia and moving along south of it, past New Zealand and across the Pacific Ocean to California. It took the swell about 2 weeks to travel so far.

*The energy in waves*

We saw that the energy in waves travels with the speed of the groups. How much energy do waves have?

We saw that a travelling wave-train can be thought of as the sum of two similar standing waves. This makes the argument easy.

One standing wave is at its maximum elevation. It contributes a pattern of elevation (and consequently, potential energy). The other standing wave at that moment is 'flat'. It contributes a pattern of water velocities, and consequently kinetic energy. But looked at separately, these energies are just the whole energy of one standing wave at different stages of its history. They must be equal.

So we need only calculate one of them and I choose the potential energy. How does the potential energy arise?

In comparison with a flat sea, water is missing over the troughs of the wave and water has appeared under the crests. It is not actually the same water, but in thinking of the potential energy we can argue as if it were. Water has been, in effect, taken out to make troughs and piled up to make crests.

Think of a column of water in the crest, height $z$ above the mean level, width $\delta x$, and of length $y$ parallel to the crest lines. It is as if

---

† This counts in two's instead of in ten's. For instance, if you have four positions at which you can punch holes, then you can count up to 15. I will show you the first eight numbers: a zero means no hole, a 1 means a hole is punched.

|  |  |  |
|---|---|---|
| 0 0 0 1 means 1 | | 0 1 0 1 means 5 |
| 0 0 1 0 means 2 | | 0 1 1 0 means 6 |
| 0 0 1 1 means 3 | | 0 1 1 1 means 7 |
| 0 1 0 0 means 4 | | 1 0 0 0 means 8 |

this water had been lifted from a similar place in the trough. It has been lifted by its own height, $z$.

Its volume is $yz\delta x$, its mass is $\rho yz\delta x$ and it has gained potential energy:
$$\rho g y z^2\, \delta x.$$

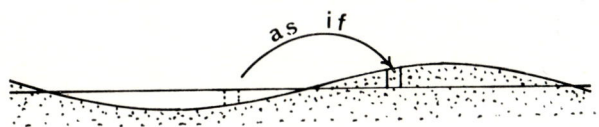

Fig. 5.11. An argument to show the energy in waves.

Now integrate this over the whole width of the crest. You can assume that the surface has a sinusoidal profile:
$$z = a \sin 2\pi x/\lambda.$$

You would get the answer:

potential energy in one wavelength
for unit length along crest $= \displaystyle\int_{x=0}^{x=\frac{1}{2}\lambda} \rho g y (a \sin 2x/\lambda)^2\, dx,$

and this works out to $\frac{1}{4}\rho g y a^2 \lambda$.

The total energy, kinetic plus potential, is just twice this, or $\frac{1}{2}\rho g y a^2 \lambda$.

Now divide by $y\lambda$ to get the energy per unit area of the surface of the sea. It is:
$$\tfrac{1}{2}g\rho a^2.$$

Energy depends just on the square of the wave amplitude.

It seems curious to me that neither the wavelength nor the frequency nor the depth of water comes into this formula. But it is the right formula.

*The energy supplied per second (the ' power ')*

Think of waves like those that reached Barbados. Suppose that in deep water their amplitude had been 1 metre, which means a height of 2 metre.

Water has a density close to 1000 kilogramme per cubic metre so the energy per square metre of sea surface would be:
$$\tfrac{1}{2}g\rho a^2 = \tfrac{1}{2} \times 9\cdot 8 \times 1000 \times 1 = 4900 \text{ joule}.$$

This energy can be thought of as being fed forward at the speed $U$, the group velocity. I find that if the waves had a period of 16 second the group velocity would be 12·5 metre per second. So multiply by this to get the energy per second sent towards the shore. It is:

60 000 watt (or joule per second).

This arrives at every metre length of shore line. If you take a kilometre of shore line (about half a mile) the figure is 60 megawatt.

*Trying to use the power of waves*

It is not surprising that on coasts where large waves are common, such as the south coast of Australia, engineers often wonder whether they could not put the energy of sea waves to some useful purpose. The difficulty in doing it is that anything that can move is likely to be swept away in a storm.

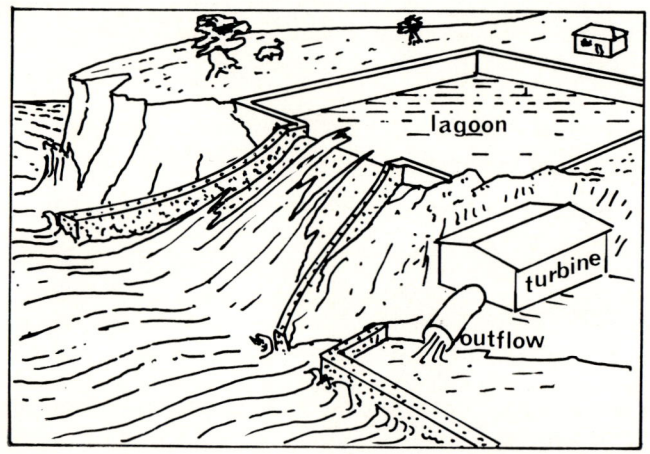

Fig. 5.12. A suggested way of making electricity from wave energy.

The most promising idea that I have heard of is that one should build a sloping wall up which the waves would run and throw water over the top into a large lagoon behind. The lagoon water could then run back to the sea in an orderly way, perhaps through turbines, and do some useful work. But I don't see how one is to build the wall both below and above the water, when the waves are likely to interrupt the work. But then, I am not a civil engineer.

*High waves at sea*

Waves at sea look big when the tops of the crests are as much as 3 to 4 metre above the water level in the troughs.

I have tried to sketch the outline of such a wave in fig. 5.13 *a*. The result is disappointing. This is because the distance from one wave crest to the next would typically be about 100 metre, and when I draw the outline it does not look at all alarming; it does not look steep. No doubt one would think differently of the real waves if one were riding in a small boat.

Waves can be still higher. A ship in a bad storm may find itself among waves 8 metre high, though this is unusual. Ships in the worst seas have reported waves 17 metre high. These are tremendous waves. I have tried to sketch the outline of one in fig. 5.13 b, assuming

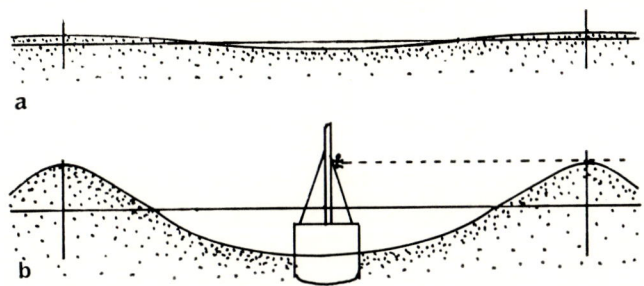

Fig. 5.13. Ordinary waves and tremendous waves. The man up the mast is estimating the wave height.

a wavelength of 100 metre as before. Even the diagram begins to look alarming now.

Real waves at sea do not have the smooth outlines one tends to draw when trying to reason about waves theoretically. Real waves are mixtures of short waves on long ones so a wave at sea is rather like a mountain range, with isolated summits and valleys and foothills. If you have read *Treasure Island* you may remember a striking description of waves as seen by Jim in the coracle. I can tell you how R. L. Stevenson came to know what waves were like. His father was an engineer who concerned himself with waves and the damage they could do to harbour walls. He would at times take a boat out to sea in order to look at waves and measure them. Young R. L. S. was taken with him to write the observations down. I have been told that young R. L. S. disliked these adventures, but it seems that he remembered what sea waves are like.

It must be difficult to take measurements when everything is changing, and nothing stays still. Here is a way of estimating wave heights. You climb up or down inside the ship (you may even need to climb the mast) until you find a place where the top of the oncoming wave just about blocks out your view of the wave tops beyond. This will tend to happen when the ship is in a trough. Then your height above the water is the wave height. Because waves differ in height it is usual to find a level at which about one wave in three blocks the view, and the others don't. The height you get then is often called the ' one-third highest wave height '. You may perhaps meet this phrase if you read elsewhere about waves.

*Internal waves*

People have found that there can be waves inside the sea as well as waves on its surface. This is because all sea water is not alike. The water may be cold, which makes it denser because water contracts as it cools. Or the water may contain a little more dissolved salt than usual. This again will make it denser. In either case the dense water tends to settle below the less dense water. As a result the sea is really layer upon layer of slightly different water, the density increasing as you look farther down.

The waves I am speaking of now are waves on these layers below the surface. Oceanographers call them ' internal waves '.

Wind can create internal waves if it blows towards a shore. I have tried to picture this in fig. 5.14. A wind tends to make the surface water move forward, so the surface water drifts towards the shore.

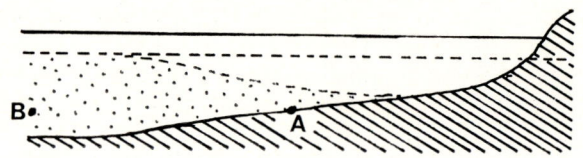

Fig. 5.14. One way in which internal waves can be started.

It must go somewhere. It piles up at first against the coast, raising the average level by a few centimetre. As the shoreward drift continues the water makes its escape by sinking and moving under the surface out to sea again.

Imagine now that the sea has a layer of slightly denser water below the less dense surface layer. The boundary between the two types of water will not be very sharp but I have drawn a horizontal broken line to indicate the approximate position that the boundary might have if there were no wind.

When the wind blows and the surface water near the shore begins to pour downwards it pushes the denser water down as well. I have shaded the diagram to show how the denser water might lie when the wind has been blowing for some time and a ' steady state ' has been reached. The denser water has been pushed down near the shore, but out at sea it has not.

This circulation under the surface is quite slow, so it is almost a hydrostatic situation, and pressure will be almost the same at equal levels, say at point A near the shore and point B out to sea. Above point A the water surface is a little higher than out at sea but, on the other hand, there is smaller depth of the denser water. The two effects combine to give no total change of pressure at point A. If the wind has raised the level of the surface water by perhaps 3 centimetre, how far must the level of the denser water have gone down?

This depends on the difference in density of the two types of water, but if the difference were only one part in a thousand I think you will agree that to get an unchanged pressure at point A, the denser water over A would need to be lower by 3000 centimetre or 30 metre. This is a big depression†.

When the wind stops blowing things tend to return to normal. The water near the shore drops to its normal level and then denser water moves slowly forward to fill up the big depression. It may overshoot the equilibrium position and reach it only after a number of slow oscillations.

Currents are another possible cause of internal waves. Waters of different types are quite likely to be drifting in different directions. For instance, the surface water may be moved along by the wind while the layer of denser water remains at rest. When two streams of fluid move past each other the boundary surface between them may be flat at first but it can be shown that the situation is unstable. I will not attempt to prove this but I will just say that the surface tends to develop corrugations that grow larger and larger. Sometimes this can be seen happening in the atmosphere. Whenever you see clouds like small parallel waves, and cirrus clouds often take this shape in the late afternoon, it is a sign of there being two streams of air, one above the clouds and one below and the layer of cloud is the boundary. The lines of cloud show the crest lines of the 'waves of instability' and it is thought that the crest lines set themselves at right angles to the relative velocity of the two streams of air.

Internal waves in the sea can be very large. It is quite common to find the denser water slowly rising and falling by as much as

† Perhaps I should work this out. Suppose that above point B there is a layer of lighter water of density $\rho_1$ and suppose that its thickness is $h_1$. Below this is denser water, density $\rho_2$, that continues down a further distance $h_2$ to the point B. For the pressure at B due to the water above it, I write:

$$\rho_1 g h_1 + \rho_2 g h_2.$$

Above point A the sea surface is a little higher, by 3 cm. Also the denser water has moved so that the top of it is lower by the distance $H$ metres that we wish to find. Consequently the thickness of lighter water above A is $(h_1 + 0.03 + H)$ metre while the thickness of denser water above point A is $(h_2 - H)$. The pressure at A is then:

$$\rho_1 g(h + 0.03 + H) + \rho_2 g(h_2 - H).$$

If the two pressures are the same I can write the difference between them as zero, and this reads:

$$\rho_1 g(0.03 + H) - \rho_2 g H = 0.$$

Rearranging this I get:

$$H = 0.03 \rho_1 / (\rho_2 - \rho_1).$$

The two densities are nearly the same. They might be (in kg m$^{-3}$):

$$\rho_1 = 1030, \quad \rho_2 = 1031,$$

which is a difference of nearly one part in a thousand. Then $H$ would be:

$$H = 0.03 \times 1030/1 = 30.9 \text{ metre}.$$

30 metre. The motion is very slow because the pressure forces that cause the motion arise from the very small difference in density between the two types of sea water. Such slow motion (perhaps one cycle in half an hour) means that there is little energy associated with them.

Internal waves do not affect surface ships. They can affect submarine ships. A submarine may be at rest at some level and then, without any change being made to its buoyancy, begin to sink farther and farther. Too great a depth can be dangerous of course, but if no action is taken, the vessel may be shortly found to be rising again. Then it begins to sink again and the oscillations continue. It is riding on an internal wave.

## Waves coming ashore

Waves grow higher as they come towards a beach. We can now see why.

Out at sea they are feeding energy towards the shore. In shallow water the wave speed, and the group speed too, is reduced. So in shallow water the waves must be higher, and have more energy per

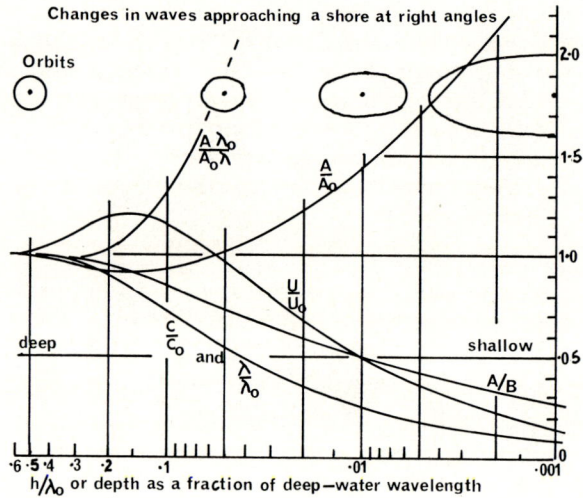

Fig. 5.15. Some calculated graphs showing how waves change when they move into shallow water.

square metre, in order to keep feeding the energy forward. If they are not high enough, energy accumulates until they are. The rule is:

$$a^2 U = \text{constant}.$$

We have not worked out a formula for the group velocity in shallow water. If you should try it, starting with the formula for the wave velocity $c$, remember that it is the frequency of the waves that must be the same at all stages in their progress.

I will do no more than show some calculated results. The diagram supposes that we start with waves whose wavelength in deep water is some value $\lambda_0$. On the horizontal axis I have arranged a scale of places where the depth of water is supposed to be different fractions of this wavelength. You can imagine the waves as coming from deep water on the left and moving into more shallow water as they advance.

The vertical scale shows just a ratio. For instance, the line $U/U_0$ shows the ratio between the group velocity at any place and its value in deep water. It grows less as the depth gets less.

As the group velocity grows less, the wave height grows greater This is indicated by the line $A/A_0$. Again, this is a fraction, the ratio of wave height at any place to wave height in deep water.

*Where to watch for far-travelled swell*

If you are interested in watching for swell from distant storms, the best place to watch for it is on the beach in the surf. The diagram I have just given shows you why.

In a boat on the open sea you might well fail to recognize the swell. It is so easily obscured by shorter steeper waves from local wind. But as the swell comes into shallow water it grows in height much more than do the shorter wind waves. At a place where the depth of water is, say, 1 metre, this is perhaps only 1/200 of the wavelength of the swell in the open sea. Then the height will have increased by a factor 1·7. But for the shorter waves the ratio $h/\lambda_0$ may be only 1/50. The same diagram shows that their height will have increased only by a factor 1·2. Consequently, if you watch the breaking waves you may well be able to pick out long crest lines appearing at regular intervals of 10, 15 or even 20 second.

Just a warning point—if you are taking holidays on the east coast of Britain, looking at the North Sea, don't expect to see swell of period 10 second or longer. You will have to be content with periods of 7 or 8 second I think. It is only on much wider oceans that the distances are enough for the longer period swell to separate itself from the rest.

*Refraction in shallow water*

On sandy beaches, and by that I really mean a beach where the water deepens only gradually as you go seaward, waves near the water's edge always succeed in lining themselves up parallel to the edge. They almost look as if they had come straight in from the open sea, no matter which way the wind may be blowing.

This happens, of course, because the shallowing water makes them bend round into this position. It is easy to make a drawing to show

how the change takes place. In fig. 5.16 I have imagined a straight coast where the depth increases regularly out to sea. The thick line represents a straight wave crest where the water is so deep that the wave is not affected by the presence of the sea-bed. The wave crest

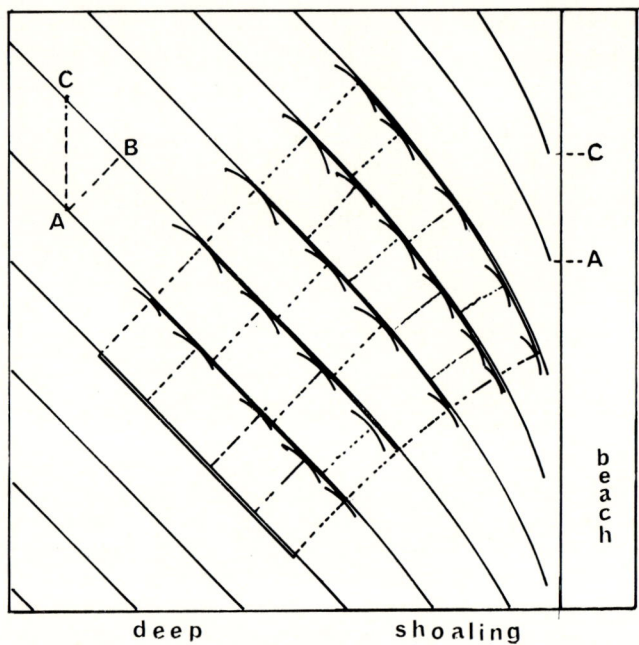

Fig. 5.16. Shoaling water makes the wave crests turn and become more nearly parallel to the beach.

should continue on towards the shore of course, but one does not know, at the start, quite how it should be drawn. We shall find this out later.

Having started with this wave crest I found the position of the next one ahead by using Huygens' construction. I took a number of points along the line and using these as centres drew circular arcs at a radius equal to the wavelength at that place. The envelope of these arcs shows the new wave crest. One goes on in this way from wave to wave, and you can see that as one goes forward into the shallow water where the wavelength is less, the crest becomes more and more refracted till finally the wave is almost parallel to the line of the beach.

On a straight beach like this the form of the wave crests should be the same everywhere along it, so the original wave crest that I took would really continue towards the beach in a curve just like the curved waves that develop from it. Indeed, the crest lines ought to be all of just the same shape, the only difference between them being that the crest line is shifted farther and farther in a direction parallel to the beach. So I have been able to extend them with thin lines that curve in just the same way as the crests given by the Huygens construction.

When the coast is straight you may be able to look at the surf and tell the direction in which the waves are moving out at sea, in deep water. First, you look at just one place on the beach and you count the number of waves arriving in, say, 2 minutes, to get the wave period. Suppose for example that it is 8 second. Then you look along the beach and you will notice, if conditions are right for this method to work, that each wave begins to break at one place and the break travels rapidly down the coast along the crest line. You should see how far the break travels along the beach in the duration of one wave period, say 8 second in the case we are imagining. You can pace out this distance reckoning 1 metre as a long step. This is the distance that the waves move parallel to the beach in one cycle, and it should be the same in the deep water as it is at the beach itself. What you have measured should be the distance AC in the diagram. Perhaps it may be 140 metre. But the wavelength is the distance AB, and you know this from the period. For a period of 8 second the wavelength in deep water is about 100 metre. You can see then that the crest lines in the deep water must in this example make an angle of 45° with the coast, because of the rule :

$$AB/AC = \sin \alpha.$$

The rule won't work on an irregular coast ; you might very well find that the distance AC in the surf turned out to be less than the deep-water wavelength. But when a coast is irregular, engineers often use a Huygens construction to find where the waves are going to go, and to judge how high the waves are likely to be at some place where they plan to put some sea wall, or railway embankment, or some such engineering work. Figure 5.17 shows such a ' refraction diagram ' in three stages. First, just the chart of the coast with its depth contours. The second map has started with a straight wave crest coming in from the open sea and the wavelet method has been used to show how the waves refract round the headland because of the way in which the water changes depth. I have omitted the depth contours here. In the third map I have taken points equally spaced along the original straight wave crest and from them I have drawn lines going along always at right angles to the wave fronts. You might call them ' wave rays '. With water waves it is fair to assume that energy is carried forward at right angles to the crests, parallel to the rays, so where the rays open out the energy is spread over a wider front and the waves are lower. One can, in this way, judge what height of wave is likely to occur at different places behind the headland, starting with some assumed height of waves in the open sea.

In this example the water off the headland is so deep that most of the energy passes it with little refraction and places round the corner of the headland are well protected. But fig. 5.17 also shows the situation when longer waves are being considered. The depth of water off the

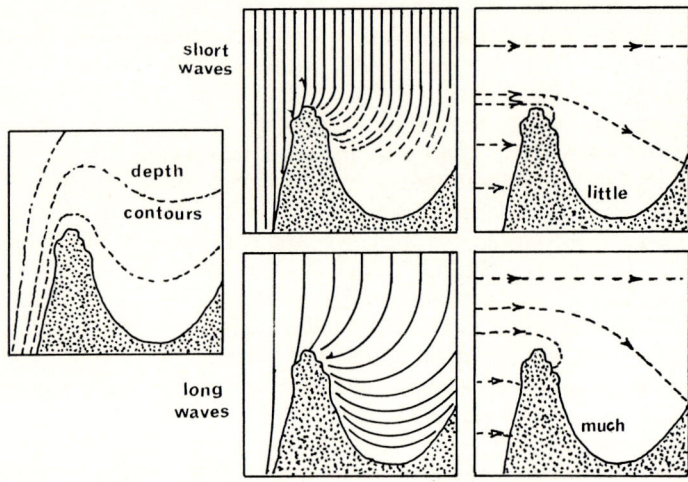

Fig. 5.17. Short waves may pass a headland and not enter the bay beyond, but longer waves will do so. The energy paths (broken lines at right angles to the wave crests) show the different widths of sea from which the waves in the bay get their energy.

headland is now a smaller fraction of the wavelength and the waves are refracted much more. Quite a lot of the incoming wave energy now reaches places round the corner.

I remember one occasion on which I noticed that on the seaward side of a headland like this the waves were rough and stormy, with a period of about 7 seconds. When I walked about a mile or so across to the other shore round the corner of the headland the sea was calmer, with waves breaking perhaps 1 metre high, but their period was 15 seconds. This was swell that had come from some distant storm. It was present, of course, on the seaward side too, but there it was obscured by the waves created by the local strong wind. The long swell had been refracted round the headland much more than the local wind waves.

*Off-shore shoals and canyons*

Refraction can have unexpected effects. Figure 5.18 shows a straight coast that would seem uniform to someone on the shore, but actually there is an underwater valley leading out to sea. You might well expect that when swell approaches such a coast, the surf would be higher near the beach in line with the deep-water channel because the deep water does not hold back the waves. But I think you will agree that the argument works the other way round. I made a Huygens construction for this coast and then drew the wave rays that are shown in the second diagram. Because the waves go forward

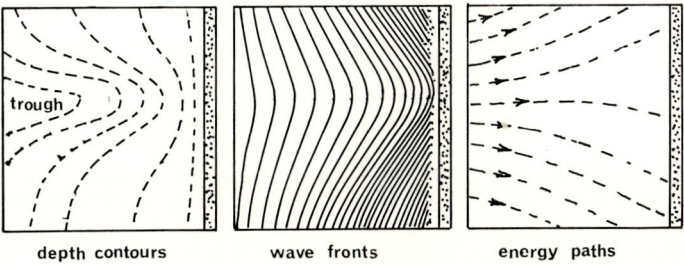

Fig. 5.18. Waves coming from the left towards a shore on the right. An underwater valley will deflect wave energy away from the beach at its head.

faster in the deeper water the wave energy is refracted towards the shallower water on either side. It would be easier to bring a boat ashore at the head of the underwater valley than at other points on the coast.

On the other hand, a patch of shallow water, an underwater hill, can focus the waves on to a place beyond it. This presumes that the waves do not break as they pass over the shoal itself; if they do, then the shoal will protect boats in the lee of it; if they don't, then the waves in the lee tend to be higher than if the shoal were not there.

*Physical arguments*

We have not seen much mathematics in the last two chapters. We have been using graphical, or pictorial or 'physical' arguments very often. I think this is the best way. It gives one good ideas of what is likely to happen. Then, if it seems necessary to do so, one makes up some mathematics to express the argument. I think you could make up some more mathematics to go with these ideas of refraction. For example, you might find a method or a formula for drawing a wave ray directly without needing to draw the wave fronts (the crest lines) first.

Physical arguments sometimes lead to incorrect conclusions, but so do mathematical arguments, too. I like physical, pictorial arguments myself because they are so quick to use and easy to remember. It seems to me that one can scarcely do without physical arguments when working in physics; without them one would not know in what direction to drive the mathematics.

# CHAPTER 6
## more about sea waves

THE study of earthquakes is called seismology (pronounced like ' size ') and seismologists have instruments that they call seismometers whose purpose is to record the motions of the ground. These instruments can be made very sensitive but there is a point beyond which it is not worth increasing the sensitivity because it would merely result in the instrument being continually in motion due to small ground motions that have nothing to do with earthquakes. These small motions of the earth are called ' microseisms '.

### Waves that shake the ground

Microseisms have many causes. Near towns or highroads they may come from machinery or from moving vehicles or from tall buildings swaying in the wind. In the country they may be caused by wind moving trees whose roots move the ground. These motions are usually rapid, 1–10 cycles a second, frequency 1–10 Hz.

Fig. 6.1. Microseisms have many causes.

At the other end of the scale there are tidal motions. If you are near the coast you will understand that when the sea tide piles up an extra foot or so of water over all the nearby sea, this extra weight pushes the sea-bed down a little and the whole country tilts a little. One can make seismometers that detect this tilt. These tidal microseisms are very slow of course. They have the period of the tides.

Intermediate in this range one finds microseisms of period between 3 and 10 second. These are often very regular. They come in groups, a few big ones, then small ones, then big again, but usually showing a very characteristic period. It may be 7 second on one day and 5 second on the next. Indeed a typical record of these microseisms looks very like a record of ocean swell and it is now thought to be ocean swell that causes these earth waves.

It was seismologists in India who first insisted that this kind of microseism arose from storms, meteorological ' depressions ', but only while these storms were over the sea. When the storm passed the coast and lay over the land the microseisms ceased. They had seen this happen too often to doubt it.

But what could be the explanation ? Perhaps it was waves striking the coast. But sometimes the storm started so far out to sea that the waves had no time to reach any coast before the microseisms began to be noticed ashore. One suggestion that was put forward was that perhaps, in storm centres, there was a tendency for the barometric pressure to fluctuate at this period of from 3 to 10 second. After all, a circular storm is rather like an inverted drain-pipe where air at sea level spirals in towards the centre and then rises up to some much greater height. Perhaps the air in such a system would tend to resonate like air in an organ pipe tends to resonate at a frequency appropriate to the length of the pipe. If the pressure at the storm centre did fluctuate in this way one could understand that as the pressure fluctuated over a wide area the load would be transmitted to the sea-bed and produce seismic oscillations that would radiate away in all directions at the usual speed of such seismic waves, about 5 kilometre per second. But no one had ever noticed such fluctuations in air pressure taking place, and, again, if it could happen when the storm was over the sea why should it stop happening when the storm crossed the coast and was over the land ?

Then a French scientist working on the West African coast, at Dakar, noticed that his seismometer began to show strong microseisms whenever heavy swell coming from storms in the North Atlantic reached the coast. Swell striking the coast produced microeisms, but he noticed an odd thing ; when the swell had a period of about 14 second the microseisms had a period of about 7 second. The period of the microseisms was half the period of the swell. It was easy to believe that heavy swell striking the coast might cause earth tremors, but what was one to think about this curious factor of 2 ?

It seemed as if waves at sea could be a source of microseisms. Yet how could waves on the surface of the deep sea affect the sea-bed 4 kilometre below ? The pressure fluctuations in the water would have dwindled to something quite negligible even at a depth of one wavelength, say 200 metre.

Then two young mathematicians working at Teddington near London came to light with an idea that seemed to be right. I will give a physical argument for it.

Waves at sea can be travelling in precisely opposite directions. One can see that this may happen when waves strike a coast, if some fraction of them are reflected back out to sea from cliffs or rocks. But it may happen too in the deep sea when a storm centre is moving. The wind direction reverses as the storm passes and new waves may then be generated, travelling quite the opposite way to the ones that were formed at first.

So think of a standing wave. First it is ' flat '. Then a quarter of a cycle later the water is standing in crests and troughs. Some of the water has gone down and some of the water has gone up. At first thought one would say that these amounts would be the same, yet on

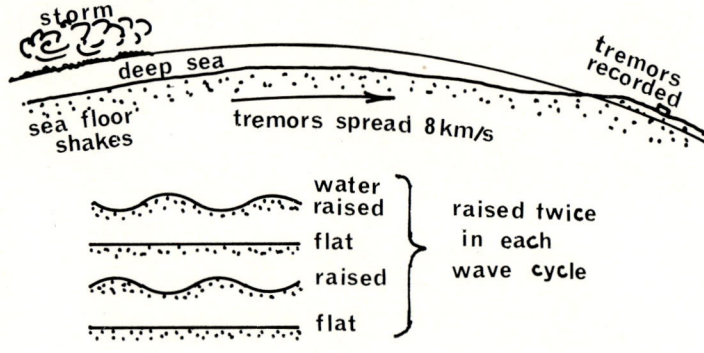

Fig. 6.2. Why the standing wave shakes the sea-bed.

second thought, they can't be. You could get the final shape just by taking water out of the troughs and moving it up into the crests. So it seems that rather more water must have gone up than has gone down. You can express it in this way; when the water stands in troughs and crests, the centre of gravity of the sea is a little higher than it was when the wave was flat.

But when the standing wave has inverted, half a cycle later, and troughs have turned into crests, the same is true again. The centre of gravity is higher once more. Twice in every wave cycle the centre of gravity of the sea rises and falls.

But this cannot go on without something feeling the reaction; you can't push one thing one way without pushing something else the other way. You need not consider the centre of mass of the entire sea of course. Just imagine a plane sketched at a depth of about one wavelength. The centre of mass of the water above this level is certainly rising and falling and the water below must produce a fluctuating force to cause the rise and fall. If this deeper water does not move, the reaction is transferred to the sea-bed even if it is 4 kilometre below†.

I hope that you find this physical argument convincing. One might almost call it obvious when it is pointed out. Yet this is not the way in which the idea first came to light. The two mathematicians were reading an article by a French mathematician who was trying to argue as exactly as he could what the behaviour of a standing wave should be. They noticed that in the formula for pressure there was one term that did not decrease with distance downward. It fluctuated at double the frequency of the standing wave. They realized that it could be this part of the pressure that was responsible for microseisms. So the idea came to light. I wish it had been my idea.

## Detecting microseisms

One of the Indian seismologists I spoke of described to me a simple arrangement for detecting microseisms. You may like to try it, but it will need to be placed in some cellar where it is less likely to be affected by casual vibrations from people moving about the building in which you keep it.

Take a wide bowl and a rather smaller flat metal can that will fit inside the bowl as in fig. 6.3. I suggest metal because you need to drill a small hole through the side of the can about halfway down. The bowl is to be filled with water so that the can is covered. A thin elastic sheet, perhaps the skin of a large rubber balloon, must be stretched across the top of the can and tied firmly in place so as to close the top completely. A metal weight is now placed on the rubber sheet. The weight must not sink so far as to rest on the bottom, it must be supported by the sheet. Now, if earth tremors tend to move the apparatus up and down, the weight tends to be left behind, but in doing so it must pump water in and out of the hole in the side.

To detect a flow of water through the hole you hang over it a small flat mirror, as light as possible. This is deflected as the water moves in and out. You detect the deflection of the mirror by a beam of light in the same way as you detect the motion of a mirror galvanometer. If the outer bowl has a glass side you can send the light

---

† Water is compressible, of course, so the water below the wave can have a vertical motion. Indeed the whole sea can resonate vertically like air in a vertical organ pipe. The sea surface will be an antinode and the sea bottom will be a node if we can disregard the motion of the sea-bed. So the total depth, say 4 kilometre, will be either one-quarter of a wavelength or 3, 5, etc. quarter wavelengths of this compression wave (not the *surface* wave now). If you take the speed of sound in water to be 1·75 kilometre per second you will find that the natural vertical resonances of the sea have periods of 9 second, 3 second, 1·8 second, etc.

The possibility of these resonances complicates the argument, but it still remains true that the bed of the deep sea will experience a uniform fluctuating pressure at double the frequency of the standing wave on the surface.

through it, but otherwise I suggest two mirrors, one tilted at 45° in the water and one tilted at 45° in the air above.

When you use an optical beam as shown, the lens A in front of the electric light bulb serves to focus the filament of the bulb on the small mirror and to get as much light on it as possible, so its focal length

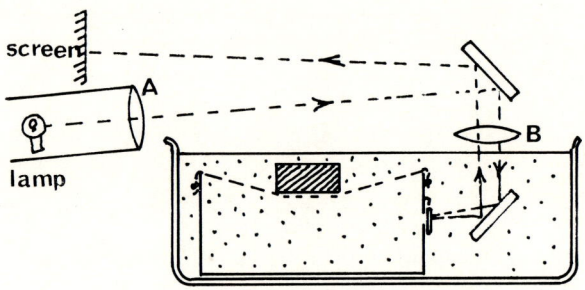

Fig. 6.3. An unorthodox seismograph (for vertical motion).

needs to be a little less than the distance of the lens from the bulb. The lens B serves to focus an image of lens A on the viewing screen ; its focal length needs to be about the same as its distance from lens A or from the screen, because the light goes through it twice.

If you wish to increase the sensitivity of the apparatus you may use a larger metal weight, but then it would be necessary to support most of the weight on a long spring so that not too much was left to be supported by the rubber sheet.

' Rip ' currents in surf

If you go bathing on a coast where there are large breakers you will have been warned that here and there along the shore, at points which vary from day to day, there are places at which the water on the shoreward side of the breakers gathers together and streams out to sea. These are called ' rip ' currents. You may be able to detect these places from the beach or from some cliff behind by seeing that the waves are rather confused or by noticing the motion of patches of foam. If the shore material is fine enough to stay in suspension you may see a streak of sandy or muddy water reaching through the surf out to sea.

I may as well pass on some advice given to me. If you find yourself being carried seaward in a rip, do not exhaust yourself by swimming shoreward against the current. Aim in a direction parallel to the coast. You will still be carried seaward at first but rip currents are not very wide, perhaps 10 metre or less. In any event, they die out beyond the zone of breakers and you can return more easily to the beach at some other place.

It seems that these currents develop because waves tend to carry surface water toward the shore. The rip current is a way in which the water can return seaward.

If you turn back to Chapter 3 you can see two reasons for expecting waves to give this slow forward drift. We imagined there the 'steady-state' situation where a wave-train is held stationary on a steady stream. The real velocities of a water particle are not quite those we

Fig. 6.4. Rip currents die out beyond the surf.

proposed. Wave activity is greater near the surface than farther down; so when the particle is above its mean level the waves give it a velocity (forward) a little greater than we proposed; when the particle is below its mean level the waves add to it a velocity backwards, with the stream, that is a little less than we proposed. So on the whole the waves give it a slight forward drift. And again, the to-and-fro horizontal motions that the waves give to the water particle tend to make it dwell longer under the wave crests than under the wave troughs. This again tends to add a forward drift.

These corrections to our argument come about, as you see, because we realize that the particle is not in quite the same position as we guessed by thinking only of its passage down the stream; the waves give it a slightly different position. The greater the wave amplitude the greater will this difference be. But if the wave amplitude is greater, then the velocities given to the particles are greater too, so the corrections we are thinking of depend in two ways on the wave amplitude. If we think of a wave-train of twice the previous amplitude, the forward drift that it produces in a particle will be four times as great (the drift is proportional to the square of the amplitude).

More correctly it is the ratio of wave amplitude to wavelength that affects forward drift (as the square of the ratio). When waves approach a beach, both effects appear. The waves grow higher and they also grow shorter in wavelength. This can mean a big increase in the forward drift. When a group of big waves approaches the breaking zone they begin to suck water from the sea behind them and to pile it up on the beach.

## The beat of the surf

When you have been standing at the edge of the sea you may have noticed that the level of the water at the edge fluctuates slowly. It is particularly noticeable where the beach slope is very gentle so that waves break well away from shore and the actual edge of the water is fairly quiet. Then the water may slowly advance several feet up the beach, remain there for a minute (say six waves) and then slowly retire again. It continues to do this in a rather random way.

One investigator working in California thought he would make measurements of this fluctuation of water level. The difficulty was, of course, to distinguish these relatively slow variations in level from the large and rapid variations due to the actual waves. He used the amusing and ingenious combination of bottles, tubes and air leaks that is shown in fig. 6.5.

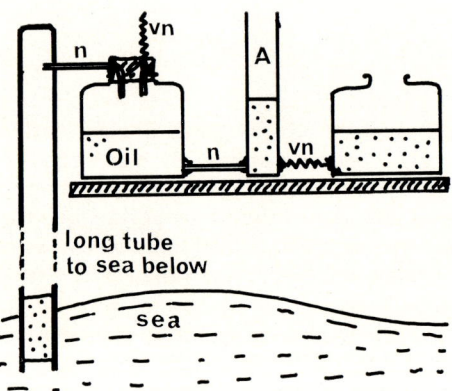

Fig. 6.5. Air trapped in the vertical tube changes pressure as the sea surface rises and falls. Watch the level of the oil in bottle A to see the ' beat of the surf '. The very narrow (vn) tubes reduce the effect of tides. The narrow tubes (n) reduce the effect of ordinary waves.

Having been the first to make measurements of this rise and fall of water level due to waves on shore he had to invent a name for it and called it ' the beat of the surf '.

I think you will agree with his explanation of surf beats. He argued that successive waves differ in height and that when swell arrives at a coast one commonly finds a group of large waves, five or six, arriving in succession and followed by a similar interval in which the waves are low. So more water would be brought shoreward by the high waves and less by the low ones, with the result that the water level near the shore would fluctuate.

Another investigator, in Britain this time, heard of the experiment on the surf beats and reasoned that if surf caused the water level at the shore to rise and fall in this way it was very likely that the fluctuation would radiate out to sea as very long, low waves. The apparatus he happened to have was a conventional 'wave recorder', sunk on the sea-bed about 1000 metre from a beach on the coast of

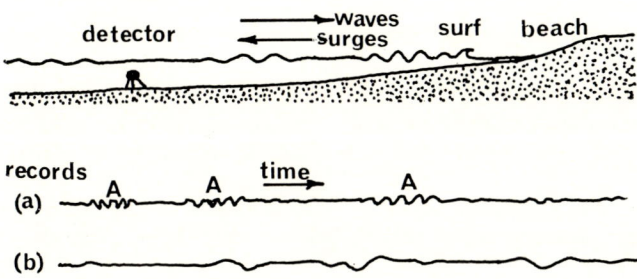

Fig. 6.6. Surges reflected out to sea from wave groups striking the shore.

Cornwall. The instrument was intended to detect the fluctuating pressure under the incoming waves, but if low waves due to 'surf beat' were travelling out to sea it should detect these too. He wanted to test whether it was doing so.

The electrical output from the wave recorder was fluctuating in the manner shown in the first curve marked (a) in the diagram. The time scale is very compressed; the individual waves typically have a period of 10 or 12 second. The whole record represents an interval of 20 minutes. The first question was whether any long-period variations were mixed up with this signal. The investigator made an electrical circuit that would let very slow fluctuations of voltage pass through it but would not pass the more rapid oscillations due to ordinary waves. He connected this 'filter' to the electrical output from the wave recorder and found that long-period irregular fluctuations did appear at the output end of the filter. They were much smaller than the fluctuations due to normal ocean waves. I have pictured them in curve (b) in the diagram but you will understand that this curve represents oscillations of a few centimetres in the water level, rather than variations of 1 or 2 metre such as ordinary waves produce. He began to make simultaneous records of both the ordinary signal and the filtered one.

The next question was whether there was any real connection between these slow undulations in water level and the arrival of groups of waves. The proposed idea was that when large ocean waves pass over the instrument at sea they proceed shoreward and reach the breaker zone in about 3 minutes. Being large waves they would raise the level of the water by the beach. This should then start a low surge travelling out to sea, and it might well reach the wave recorder after another 2 minutes.

The curves ($a$) and ($b$) that I have sketched were simultaneous records of the incoming waves and of the slow surges and there is some suggestion of a correspondence. You can see that the groups of large waves that I have marked A A A are followed about 5 minutes later on curve ($b$) by oscillations that could be the outgoing surge. However, the agreement was not always so good and the agreement might be accidental. Even if there were a real correspondence it might involve some other time delay than 5 minutes. It seemed best to make a 'lag correlogram'.

## The idea of a correlation coefficient

To explain what I mean by a 'correlogram' I had best go back to an occasion on which I myself first found a use for the idea of 'correlation'.

The matter had nothing to do with sea waves. It concerned a method of telling when a ship was passing into a harbour or out of it, perhaps at night or in fog. Ships are made of iron, and iron, even if you do not try to make it into a magnet, inevitably becomes

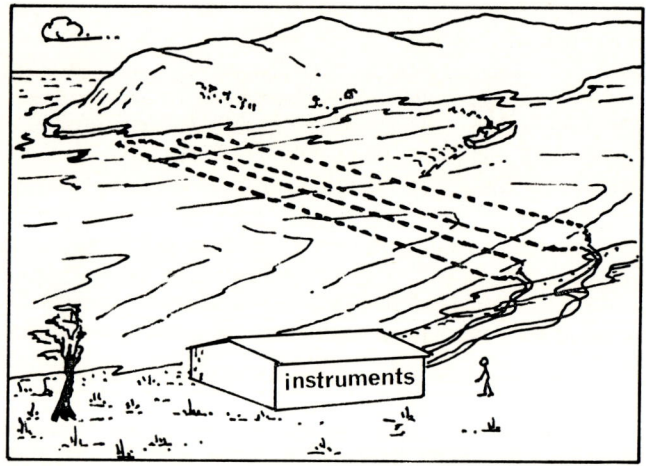

Fig. 6.7. A pair of underwater 'loops' across a harbour entrance.

magnetised by induction in the Earth's field; this method used the fact that ships are magnetized. People had laid a long loop of insulated electric cable across the harbour entrance on the sea bottom, with the idea that as a ship passed over it, an e.m.f. would be produced in the loop and that this could be detected by connecting the shore ends of the loop to a galvanometer. If the galvanometer deflected, it showed that a ship was passing.

Actually, two loops of cable were used, laid a little distance apart. The reason was that even when no ships are passing, a galvanometer connected to a single large loop of wire keeps continually deflecting, one way or the other because the magnetic field of the Earth continually varies†. But by using two loops and connecting them in opposite ways to the galvanometer, these fluctuations due to the variation of the Earth's field might be balanced out, while the effect of the ship, coming first on one loop and then on the other, would be detected. I have sketched three curves in the diagram to illustrate what was supposed to happen in principle. The first curve represents

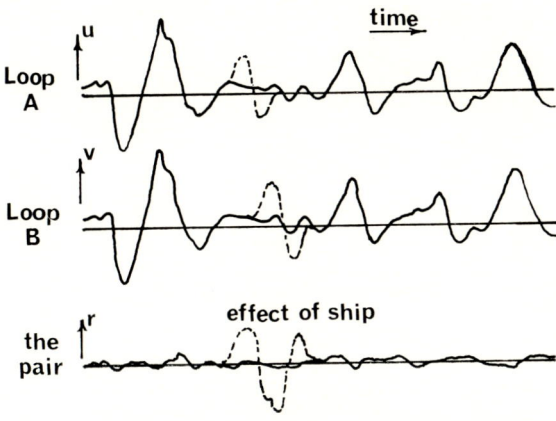

Fig. 6.8. The very small electrical signal recorded from loops A and B separately and from the pair when suitably connected in opposition.

the history of the signal from the first loop during an interval of 2 hours, the second being the signal from the second loop. You can see that the variations are nearly the same but that from the second loop is perhaps a little larger, the loop being a little larger in area. It was possible to make allowance for this difference by putting some resistance in series with the larger loop. Then when the two were both connected to the galvanometer in parallel opposition it would record a signal somewhat like the third curve.

† Sometimes it is long slow variations, sometimes oscillations almost like sea waves or microseisms that persist for hours. You could make this experiment yourself if you could get a galvanometer and some hundreds of yards of thin insulated wire to lay out in a loop round a field, or perhaps round a building and high enough to be out of people's reach.

I have shown in broken lines the extra signal one might get as a result of a ship passing. It would not be recognized on either loop alone, among all the other perturbations, but on the output from the balanced pair it stands out clearly.

But my duty was only to find the best balance point for the loops, to decide on the right value of resistance to use and to leave it like that. So that is what I did, but it was very tedious. One watched a history like that in the diagram unfold slowly in the 2 hours. Then one would try another 2 hours with a different setting of the resistance and if it seemed to make things worse, try another 2 hours with the resistance adjusted in the other sense. This went on for days.

One does what one is asked in scientific work, but it is advisable from time to time to sit back and think. The thought that struck me did not help me in the actual business of balancing the loops but it gave me a new point of view. It struck me that if one were given the output of two separate loops, say the curves ($a$) and ($b$) covering 2 hours, then it should be possible, in principle at least, to decide what the balance point should be without having continually to repeat the process and find the best balance by trial and error. All the necessary information was really present in the two curves and there should be a rational way of arguing about it.

You may very well say that the idea is obvious. Still it was new to me and I thought I might try to write it down as algebra. I wrote $u$ for the signal from the first loop and $v$ for the signal from the second. The signal that was left after I had put the two in opposition I called the 'residue' $r$. I had used a resistance to select only a certain fraction of the second signal, say a fraction $k$, so what I had done seemed to be represented by the statement:

$$r(t) = u(t) - kv(t).$$

I have written $t$ in brackets after each voltage to indicate the time since the start of the record and to show that the $u$, $v$ and $r$ voltages all referred to the same moment whatever it might be.

Now, I had been adjusting the fraction $k$ by means of the resistance to try to make $r$ 'as small as possible'. How should I represent this? What did I mean by 'small'; how did I judge when $r$ was small? The signal $r$ was a fluctuating thing, sometimes zero and sometimes not, either positive or negative. What I should be thinking of was some kind of average of $r$ throughout the 2 hours.

But it would not be sensible to take the ordinary average of $r$ as an indication of 'size'. One could very well have a case where the average value of $r$ was zero and yet it might have large positive and negative excursions at different times.

I recalled that I had somewhere heard of people averaging the square of a quantity. This seemed a promising idea because the square is always positive and if I found that the average of the square was small, then $r$ could never be large (or if it were ever large it could not remain so for long). So what I needed to think of was $r^2$. This works out to be (I will omit the $t$ now):

$$r^2 = u^2 - 2kuv + k^2v^2.$$

Then the average of $r^2$ would be the sum of the averages of the things on the right-hand side. I could denote an average by writing a line over the symbols so:

$$\overline{r^2} = \overline{u^2} - 2k\overline{uv} + k^2\overline{v^2}.$$

This was what I could reasonably call the 'size' of the residue I was left with.

Now I had been adjusting the fraction $k$ by trial and error so as to make the residue small. It seemed reasonable to represent this by calculus, differentiating with respect to $k$ and finding when $\overline{r^2}$ was least. The mathematics gives:

$$d\overline{r^2}/dk = 0 - 2\overline{uv} + 2k\overline{v^2}.$$

This will be zero at the best setting of $k$ so evidently we have an expression for the optimum value of $k$:

$$k(\text{opt}) = \overline{uv}/\overline{v^2}.$$

Given histories of $u$ and $v$ one could calculate this optimum value of the ratio $k$ without actually doing the trial-and-error experiment.

Now put this into the expression for $\overline{r^2}$ and you have the least value of $\overline{r^2}$ that could be achieved:

$$\overline{r^2}(\text{minimum}) = \overline{u^2} - 2(\overline{uv})^2/\overline{v^2} + (\overline{uv})^2/\overline{v^2}$$
$$= \overline{u^2} - (\overline{uv})^2/\overline{v^2}.$$

Finally, what concerns us is not the actual 'size' of the residue, but its relation to the 'size' of the original signal from a single loop. This ratio is:

$$\overline{r^2}/\overline{v^2} = 1 - (\overline{uv})^2/\overline{u^2}\,\overline{v^2}.$$

So it is true, as we said at the start, that all the information about the best ratio $k$, and the extent to which it is possible to reduce the fluctuations by proper balancing, is all inherent in the curves showing the histories of $u$ and $v$. The degree of balance you can get is evidently shown by the following curious number that I will call $\rho$ (Greek 'rho'):

$$\rho = \overline{uv}/(\overline{u^2}\cdot\overline{v^2})^{1/2}.$$

I have written it in this way because it then has the same sign as the ratio $k$, which would need to be negative if I had quantities $u$ and $v$ that fluctuated in opposite senses. The sign of $\rho$ tells me whether the quantities fluctuate in the same sense or in the opposite sense.

Given two histories of $u$ and $v$, if $\rho$ turns out to be zero, then no reduction is possible by balancing. Evidently this happens when $u$ and $v$ fluctuate in a way that is quite unrelated. But if $\rho$ turns out to be near to 1 (or to $-1$), then the two quantities $u$ and $v$ must fluctuate in almost exactly the same way because a very small residue is possible.

I was delighted with this curious and interesting number, $\rho$. I called it to myself a ' coefficient of similarity ' because that was what it seemed to be. It was a numerical measure of the resemblance between the fluctuations in the two quantities. Then someone pointed out to me that the number $\rho$ had been invented 40 years before by Professor Karl Pearson and was called the ' correlation coefficient '. What a disappointment! Yet it was a piece of good luck, too, for it made me feel that I quite understood what a correlation coefficient meant. At least it has provided me with a way of telling you what it means.

*Back to surf beats*

I left off at the point where the investigator had decided to make a ' correlogram ' to compare the fluctuations in the long-period surf beats with the fluctuations in the amplitude of the ocean swell. First

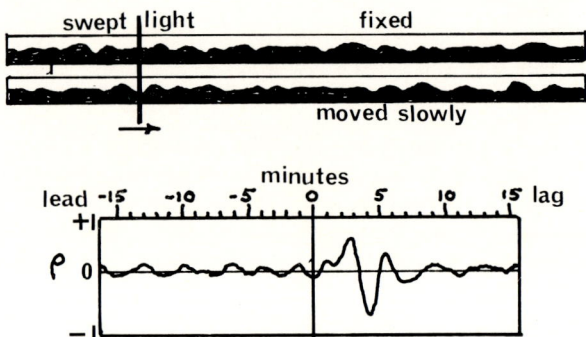

Fig. 6.9. The ' lag correlogram ' between records of wave groups and of surges. Apparently the surge tends to arrive $4\frac{1}{2}$ minutes after the wave group passes.

he took the ordinary record showing the swell and drew on it a curve touching all the tops of the swell. This curve then represented the amplitude of the swell, showing how the amplitude varied with time†.

† Since the forward drift of water depends on the square of the wave amplitude, perhaps he would have done better to make a curve representing (amplitude)$^2$. But I think he just drew the amplitude curve, as I have said. Anyway, it worked, as you will see.

Now he had two curves to compare. I have sketched them again but this time I have drawn them as profiles of black on white, for this is the way in which he actually made them.

If he had had a digital computer I imagine that he would have measured off the heights of the two curves at the large number of instants, say every 10 seconds, throughout the history, and given all the numbers to the computer and made it work out the answer. But he didn't have one, which is why he drew the curves as profiles. He intended to focus a strip of light across the painted profile and sweep it along the length of the record, using a photocell to pick up the reflected light and so give an electrical signal that was a speeded-up imitation of what the record showed. If he did this with both records at the same time he could make an electrical circuit to imitate the calculation. One calls this an 'analogue' method of making a calculation because one is not working with actual numbers but with quantities such as light intensity or electric current whose variations imitate those of the quantities one would calculate if one were using numbers.

Now he saw a difficulty. He could get two electrical voltages varying with time in imitation of the two profiles. I will call them $u(t)$ and $v(t)$. To calculate the correlation coefficient he would somehow need to multiply them together to get the average product that I wrote as $\overline{vu}$. He couldn't see how he could easily make an electrical circuit to give an output that was proportional to the product of two voltages. On the other hand, he could see how to get something that was approximately the square of just one voltage. So he used the following trick, that you may find useful too, for even when you work with a slide-rule it is easier to read off the square of a number than to multiply two numbers together.

He put both records together in front of the photocell. Its output was then the sum:

$$u + v.$$

Then he used his electrical circuit to give a voltage equal to the square of this, which is:

$$u^2 + 2uv + v^2.$$

Then he fed this voltage to another circuit that would smooth out the fluctuations in time, and give an average value, which I would write as:

$$\overline{u^2} + \overline{2uv} + \overline{v^2}.$$

I think you can see, now, how he got the value of the correlation coefficient. He used just the first profile by itself and measured $\overline{u^2}$. The second profile, alone, gave him $\overline{v^2}$. Having got these three numbers he could easily work out the number for the correlation coefficient which is:

$$\rho = \overline{uv}/(\overline{u^2}\cdot\overline{v^2})^{1/2}.$$

You will remember, however, that this investigator wanted to find if the long-period surge hit the wave-recording instrument about 5 minutes after the passage of the particular wave group that produced the surge. What he needed to do was to displace the profile representing the surges by a distance equivalent to 5 minutes of 'real time'. Then the analysing apparatus would be comparing the amplitude of the swell at any instant with the surge seen 5 minutes later. Indeed he thought he had better try all kinds of time shifts to make sure that the kind of agreement got by a 5-minute shift was actually much better than the agreement got by any other shift. So, having set up the two records, with the light beam repeatedly sweeping across them and the electrical circuit producing steady output voltage proportional to:

$$\overline{u^2 + 2uv + v^2},$$

he began very slowly to displace the surge record relative to the swell record. Indeed, he began with a displacement corresponding to 15 minutes and slowly reduced to zero and continued till the displacement corresponded to a time lag of 15 minutes in the opposite sense. The steady output was being recorded all the time by a pen on moving paper and as the displacement changed, the steady output changed too. The line that the pen drew is pictured in the diagram. It really represents how the average product, $\overline{uv}$, varied, because $\overline{u^2}$ and $\overline{v^2}$ would remain much the same. I have marked on the diagram a scale that denotes the value of the correlation coefficient. You can see that the greatest correlation comes with a delay of $4\frac{1}{2}$ minutes. You can see that it is negative. It means that each group of high waves tends to produce a *depression* in the water surface that travels out to sea.

The investigator had not expected this. He had thought (as I argued at first) that it would be an elevation. Yet you can see how it might work. When a group of high waves gets to the surf zone it begins to move water rapidly towards the shore. It must leave a depression in the sea behind it. It is this depression that moves out to sea as a surge. The curve suggests there may be a smaller elevation both preceding it and following it.

The record shows smaller correlations both positive and negative at other time delays, even for time delays in the reverse sense. A mathematician would say that these were 'statistical errors', meaning that the records had not included a sufficient number of wave groups. After all, if you toss a coin ten times you don't always get just five heads and five tails even though it seems reasonable to say that the two are equally likely to occur. You need to toss it a thousand times before you can rely on the ratio of heads to tails being almost exactly the ' right ' answer.

The smaller wobbles in the curve probably represent accidental variations, which would be quite different if the experiment were repeated with another wave record. But the big variation at a delay of $4\frac{1}{2}$ minutes is too big to be thought an accident. It seems that the wave groups striking the shore really did make low surges that travelled back out to sea.

# CHAPTER 7
## the wake of a ship

WHEN a ship or boat begins to travel forward at a constant velocity, a curious pattern of waves gradually develops in its rear. If you are on the boat you see clearly that all the wave pattern keeps pace with the boat. The pattern grows as the boat travels forward but it grows by new waves making their appearance at the rear of the pattern, the place farthest from the boat. The waves, once they are formed, never change. It is as if the boat were trailing the long pattern behind it, yet if the boat is brought to rest, the waves in this 'wake' continue going forward; there is no mechanical connection between them and the boat, once they have been formed.

*How slow waves keep up with the ship*

Some of the waves in the pattern have quite a short wavelength; they should travel slowly. Then how do they keep pace with the ship?

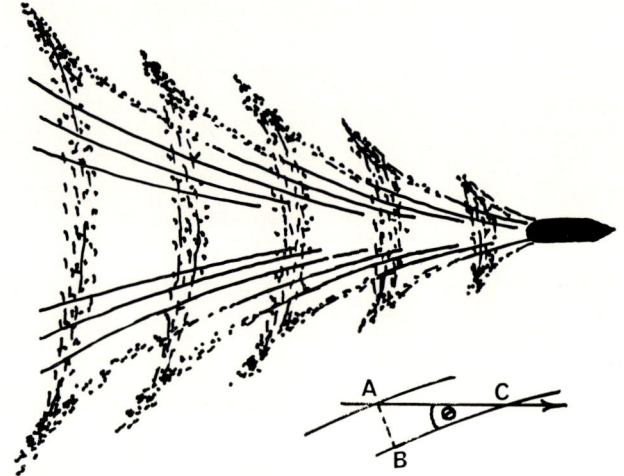

Fig. 7.1. How the slow short waves keep up with the ship.

They keep pace because their crest lines lie at an angle. We usually think of a wave as travelling in the direction at right angles to its crests. We look at the distance AB in the diagram and say the waves travel this distance in one wave cycle.

But someone in the ship prefers to look at points A and C. If this is the distance that the ship has gone in the same time, he says that the waves are keeping pace with him.

You could even write a relation between the speed of the ship, $V$, the speed of the waves, $c$, and the angle of their crests, $\theta$. It would be:
$$c/V = \mathrm{AB/AC} = \sin \theta.$$

*How does the ship disturb the water?*

The diagram pictures a ship in two successive positions. Where the ship narrows towards the bow, water is pushed outwards as the ship advances. At the stern, water is dragged in to fill the space vacated by the ship.

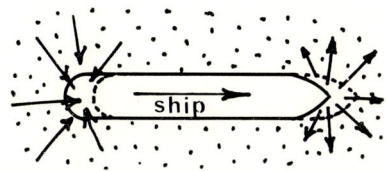

Fig. 7.2. The ship moves the water.

So the motion of the ship sets up one process near the bow, and a very similar but opposite process near the stern.

The action of the bow depends, of course, on its shape. Yet the wakes caused by different bows tend to look much alike. There are two reasons for this. One is that all bow shapes have one action in common; they all displace water away from the bow as the ship moves forward to occupy the space. The other reason is that where the wake has become very wide the pattern is produced by waves that have spread outwards from the original disturbance. The form of the pattern is largely determined by the way in which the spreading takes place.

In what follows, I am thinking of the action of the bow as a succession of small disturbances rather than a continuous action. I shall suppose that waves spread out in circular fashion from each small disturbance. Then the sum of all these spreading patterns, if I can reason what it should be like, will be the same as the wake of the moving bow so long as the disturbances are made sufficiently small and frequent.

The first step is to consider how waves spread out from some sudden brief central disturbance in water.

*Waves from a boulder dropped in a lake*

Imagine that the boulder is held half immersed in the water and then is released. As it sinks, waves will spread out from it as circles.

The problem as I have stated it would probably be too difficult for a mathematician to solve exactly; he would prefer to imagine not a stone half immersed but the water itself raised up locally into a small hump and then released from rest. This gives him a problem in which there is only one medium, the water, instead of two, the water and the stone.

The mathematician would then think of the hump of water as being the sum of many different patterns of circular standing waves (the $J_0$ patterns if the hump were quite symmetrical). I have made

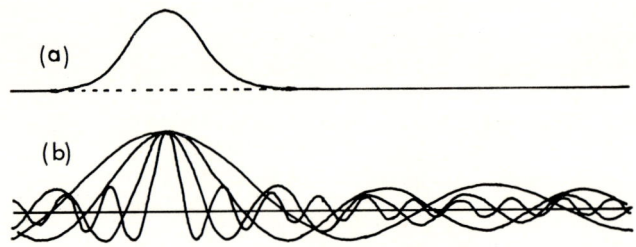

Fig. 7.3. How a mathematician would argue what becomes of a hump in water.

a diagram to show the idea. It may seem surprising that one can add together a very large number of wavy patterns and get only a single central hump, but it works if you choose the patterns properly.

The mathematician now points out that these separate circular patterns are standing waves that oscillate, and they oscillate at different rates. He chooses some future instant, finds what each separate standing wave will be like (some will have inverted and some may be ' flat ') and uses a mathematical trick to add them together again. They no longer give a central hump, of course, for they have all changed. They give some kind of circular pattern of waves and the mathematician points to this and says it is the new shape the water surface would have at that later time.

I shall not work out the details of the ' superposition ' method. Instead I will suggest a way of thinking about the problem that uses the idea of group velocity.

*How the groups travel*

All the energy that is later carried outwards, by waves spreading from the central disturbance, was originally at the central region, where the surface was elevated or the water was in motion.

I shall think of this energy as being carried out by wave groups. There will be a great variety of them, characterized by different frequencies in their waves, and they move out at different speeds from the centre. If I notice after a time interval $t$ that a group of waves is at a radial distance $r$, I shall say that this wave group is travelling with the group velocity given by†:

$$U = r/t.$$

Fig. 7.4. Longer waves travel more quickly.

## How the crests travel

I shall assume that the various wave crests travel at twice the speed of the the group to which they belong. This would be true if the wave crests were long and straight (and assuming deep water). Actually the crests are curved, being circles. It can be shown, however, that the approximation is quite a good one.

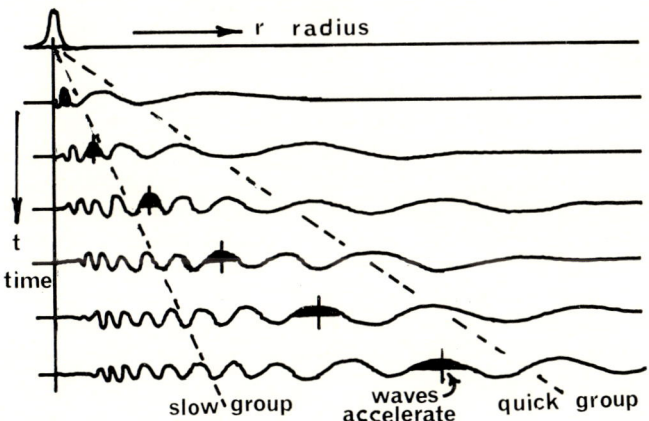

Fig. 7.5. Now we see that individual wave crests accelerate as they move outward.

† This is an approximation. The wave crests are curved, and the superposition method shows that the group speed is slightly greater if the crest lines are curved. However, the difference is very small once the waves have travelled more than a wavelength or so from the centre. You may have noticed in the $J_0$ patterns that the crests near the middle are more widely spaced than elsewhere.

121

Then I will imagine myself looking at one particular wave crest and watching it as it travels along.

At a particular instant, time $t$ after the start, this wave crest may be at a radial distance $r$. At that moment therefore it belongs to a wave group that has travelled a distance $r$ in a time $t$, and since I assume that the wave group travels at a constant speed, this speed is $r/t$. This is the 'group velocity' of the kind of wave I am looking at.

But the wave crest travels at twice the speed of the group. Therefore this particular wave crest at this particular moment has a speed $2r/t$.

Now although it seems odd, I cannot assume that the crest will keep the same speed as it travels outward. However, as time passes and $t$ increases, the distance $r$ of the wave crest will increase in some way and it is certainly right to say that the speed of the wave crest is $dr/dt$. Further, the speed of the crest must always be equal to $2r/t$ as we saw above.

Consequently I can write an equation that relates radial distance $r$ and the time $t$. It is:
$$dr/dt = 2r/t.$$
I think you may be able to show that the rule that suits this equation is:
$$r = At^2,$$
where $A$ is some number that does not vary with time†. Perhaps I would need to pick a different number to suit different wave crests in the spreading pattern.

It follows that the wave crest does not travel at a constant speed. Apparently the radial distance is proportional to the square of the time; each wave crest travels more and more rapidly as it moves outward.

This seems an unexpected conclusion. You can imagine, however, that as a wave crest travels forward it moves out of its group and becomes one of the waves belonging to the group next ahead. It travels through this and enters a group still farther ahead, and each time it does so it takes on the appropriate wavelength and speed. It continues moving into regions where the group speed, and consequently the wave speed are higher than they were before. So the crest accelerates.

Perhaps you will make an experiment and see whether this happens.

*The wake of the bow*

We imagine the bow as creating a succession of similar disturbances as it travels. Each of these events would, by itself, give rise to a

---

† Even though you might not know how to guess this solution in the first place, you can certainly check that it works. If you differentiate it to find $dr/dt$ you get $2At$ and this is in fact just the same as $2r/t$.

pattern of circular waves spreading outwards. We need only to add these patterns to give ourselves a picture of the complete bow-wake. I have pictured the idea in the diagram. The bow happens to have reached the position A, but the disturbance there has just occurred and waves have had no time to spread from it. But a similar disturbance occurred earlier when the bow was at position B and I have

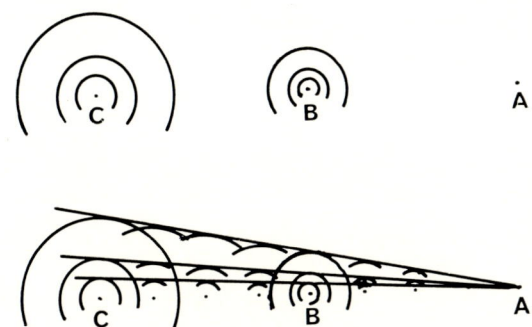

Fig. 7.6. Adding up wavelets from successive disturbances.

shown three crest lines that have spread from it. A still earlier disturbance took place at position C and I have drawn the corresponding three crest lines arising from this disturbance; since more time has elapsed, the waves from this disturbance have spread out farther.

How will these wave patterns combine? You must remember that the patterns I have shown arising from B and from C are representatives of a whole sequence of disturbances that took place all along the line of progress. Waves will spread out from all these intermediate disturbances. In the second diagram I have made the picture more complete by putting in the crest lines that have spread out from more points along the path. This diagram may suggest to you that these patterns of circular waves will combine to give three crest lines of quite a different shape, in fact the three lines that touch all the families of circles. One would call them the 'envelopes' of these families. It is the same argument as one uses with Huygens' wavelets†.

† Have you ever felt suspicious of Huygens' construction?
One starts with, perhaps, a straight wave front AA, and knowing that the wave will travel some distance $r$ in a time $t$ one draws circles with this radius. The envelope is a line whose distance is $r$ ahead of the previous one. This is obviously the right position for the new wave front. It has travelled a distance $r$ in a time $t$, as we assumed it did. So the construction gives the right answer.

But one feels unhappy about the argument if one speaks of wavelets. It is really rather complicated, of course, because the circles we draw represent only the crest lines of waves spreading from each point and we know that there will

Let us make the construction more accurately. In the diagram I marked 12 representative points equally spaced along the track and I numbered them 0 to 11. The bow is supposed to be at the point marked 0. Waves have had no opportunity to spread from this point yet, but they will have spread out from all the rest. Consider now one particular wave crest, the corresponding one in all the spreading

---

be wave troughs ahead of, and behind, each crest. When we think of the total effect at some point Q it is true that the wave from point P just puts its wave crest there, but waves from other points don't put crests at Q; some may lag so far behind that they put troughs at Q. Adding up all these crests, troughs and intermediates looks difficult. But let's skip that difficulty and

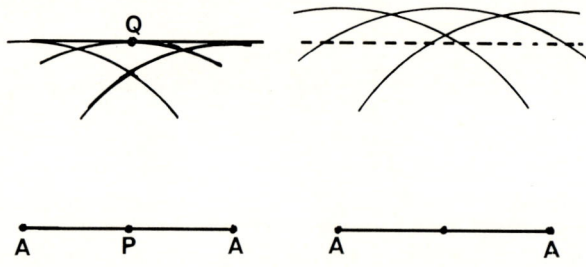

think of a simpler argument. We have many wave-trains that reach the point Q and only the one from P actually puts a crest there. *Every one of the other wave-trains puts the crest farther back.* Surely it is impossible for the combination of them all to give a wave with its crest line at Q on the envelope. Surely the combination of all these spreading wave-trains should be a crest not as far forward as the envelope.

I won't go into the business of adding up all the crests and troughs but I will say that the above complaint is justified. If we have drawn the spreading wavelets correctly, then their sum is a wave crest that is not the envelope but a line somewhat behind it.

Yet we know that the envelope is the right place for the wave front; we *assumed* at the start that the wave travelled that distance in the time interval considered. What can be wrong?

The answer is that we did not draw the circles correctly. A plane wave may travel at a certain speed but sharply curved waves travel more quickly. We have seen something like this when looking at the circular waves in a tea-cup. The wave near the centre was somewhat wider than those farther out.

So what we should have done was to draw the circles at a radius rather greater than *r*. The circles would then extend farther. Having done this we would use the right argument for adding them and find that they combined into a wave crest that was not the envelope of the circles but was somewhat behind it as the second picture shows.

In fact it would be the line through Q, which we know is the right position, and which we could have got by the usual construction which makes two wrong arguments and ends up with the right answer. This is curious, isn't it?

In the text I draw circles in the usual way and take the envelope. The result is so nearly right as to be not worth trying to correct.

patterns. Assuming that the bow has moved along the track at a constant speed, the time intervals that have elapsed since the disturbances occurred at the points 1 to 11 are just proportional to these numbers. Then the radial distances that this particular wave crest has assumed will be proportional to the squares of these numbers. For the purpose of making the diagram I have assumed that the wave

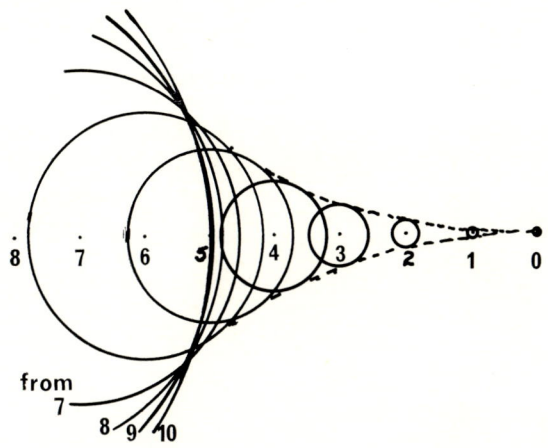

Fig. 7.7. Constructing the wake wave caused by a typical wavelet in successive disturbances. It has an arrow-head shape.

from point 1 has attained a radius just 1/20 of the spacing between the points on the track. This is not a restriction; one could always make it true by a suitable choice of the spacing. Then the wave from point 2 will have spread to a radius 4/20 of the spacing while that from point 3 will have spread to a radius 9/20 and so on.

I have drawn these circular arcs and you can see that the envelope begins as a V-shaped wave. But this V-shaped wave does not continue backwards indefinitely far. The wave from position 8 is on the V but the wave from position 9 is not. The envelope has now turned back on itself. The envelope makes a closed arrow-head shape and the wave from point 10 just touches it where it crosses the track of the bow. This is the complete envelope. The wave crest spreading from point 11 does not intersect with the wave from point 10, nor with the wave from point 12, because once we look beyond point 10 the spreading wave is travelling more rapidly than the ship.

I think you will agree that where the envelope folds back on itself, at the corners of the arrow-head, the combined wave is likely to be bigger than elsewhere on the arrow-head.

If you join the corners of the arrow to its point you will find that the angle between the lines is rather less than 40°.

There are many wave crests of course in the circular pattern that spreads from a disturbance. Each gives rise to its own arrow-head wave in the wake of the boat. In the diagram I have combined many arrow-heads to show the complete pattern of the wake of the bow. They are all the same shape. They differ merely in size. I chose

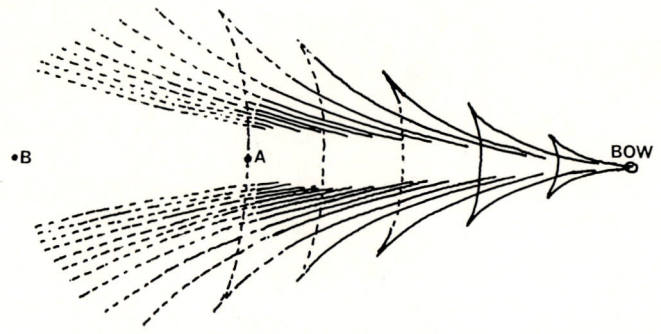

Fig. 7.8. The arrow-heads from all the wavelets that spread from a disturbance combine to show the wake pattern made by the bow of the ship.

the sizes of the arrow-heads so that the transverse crest lines, those that cross the track of the ship, are all equally spaced. You can see why this must be so. These crests form a wave-train travelling forward with the ship. There will be a definite wavelength that gives this wave-train a speed just equal to the speed of the ship, whatever that may be. The waves must have this separation.

In this diagram I have drawn part of the pattern as full lines and part as broken lines. The part shown as full lines would be present if the bow had suddenly started from rest at point A. The part shown as broken lines would be missing in that case because this part of the wake would be due to disturbances produced on the track previous to the point A. If the bow began its motion at point A, then these previous disturbances would not have occurred. The part of the wake shown as broken lines arises from disturbances between points B and A. How can I justify these statements? You can argue as follows. If you look at any part of the wake and want to know where and when it was originally formed, just draw a line at right angles to the wave crests and see where that line cuts the track of the bow. This is the position of the disturbance that produced that part of the wake.

You can see now why it is that the wake pattern grows longer as the ship travels farther†.

You can also see that if the bow had started suddenly from B and moved steadily to A and had then stopped, the wake would be the part shown as broken lines; the part shown as full lines would be missing. Some of the wake is ahead of the boat. This seems odd, but of course we are thinking of what the wake looks like some time after the boat has stopped at position A, the time at which the bow would have reached point 0 if it had continued moving at the same speed. The boat has stopped at point A and some of the wake has run ahead of it.

*The wake of the stern*

I have been speaking so far of the wake produced by the bow of the boat as it continually pushed water aside. It is obvious, however, that the stern of the boat must also produce a disturbance in the water. The action there is that water is being continually drawn inwards to fill the space left by the stern as it advances. So the stern too must produce a wake, with this difference, that crests and troughs are interchanged in position. The wakes of the bow and of the stern should be alike except that elevations in one will correspond to depressions in the other.

So the wake left by an actual boat is the sum of these two wakes. You can imagine what may happen. If the crests due to the bow happen to coincide with troughs due to the stern, then the waves in the wake may be quite low. On the other hand, when crests coincide with crests the waves in the combined wake may be high.

† In shallow water there is a limiting speed for waves. Each disturbance made by the bow would spread out with a leading wave crest travelling at this limiting speed. If you imagine the boat itself travelling at just this speed, then the leading waves from all the past disturbances just keep up with the boat and combine into a bow wave that grows higher as the boat travels farther.

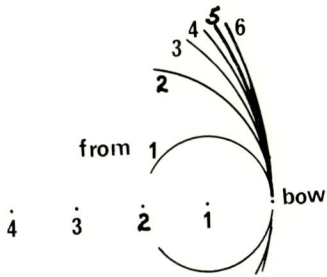

Similarly the disturbances made by the stern combine to make a trough that lies under the stern. The boat is trying to climb a hill that grows steeper with time. I suggested this in fig. 1.16 of Chapter 1.

## The wave-making resistance

A given boat has certain speeds at which it produces large transverse waves in its wake, and other speeds, intermediate between the first set of speeds, at which the transverse waves it produces are quite low. The diagrams show how this happens. They represent not the whole pattern of the wake but merely profiles of the water surface along the line of travel of the boat. The full line indicates the

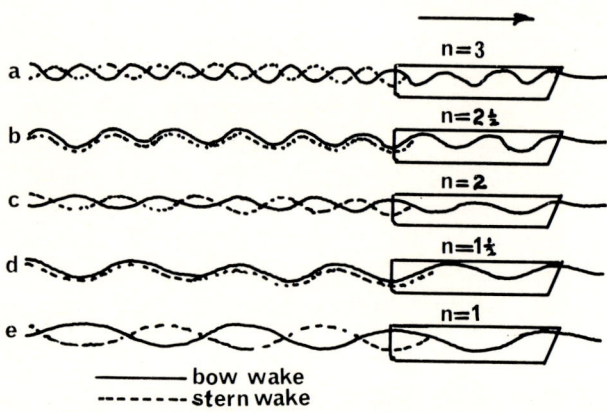

Fig. 7.9. The stern makes a wake too, but with crests and troughs interchanged. Here we look at just the section of the wake along the track of the ship and see the transverse waves (the back of the arrow-heads).

profile that the bow alone would produce, while the broken line is the profile that would be due to the stern alone. When the two patterns are out of step they combine to give only a low wake. When they agree in position the wake waves would be large. The first diagram represents a low speed, and the successive diagrams refer to successively greater speeds. You can see that large wakes and small wakes alternate.

Indeed, you will see that we should expect the wake waves to be small whenever the effective length of the boat is just equal to a whole number of wavelengths of the transverse waves. You could make a formula for calculating these speeds. If you write $L_0$ for the effective length of the boat†, then the condition is that $L_0$ is to be some whole-number multiple of the wavelength $\lambda$ of the transverse waves. We

† One might say that for the purpose of this argument the effective position of the bow is halfway along the under-water tapering portion at the front end, and that the stern is effectively halfway along the under-water tapering portion at the rear end. The effective length is the distance between this effective bow and stern. It will be less than the total length of the boat and might even be as little as one-half of it, depending on the form of the hull.

128

also have the wave formula relating speed $V$ to wavelength $\lambda$:
$$L_0 = n\lambda \quad V^2 = g\lambda/2\pi,$$
So for speeds that give only a low wake we have the formula:
$$V^2 = gL_0/2\pi n \quad (n = 1, 2, 3 \ldots).$$
Large wakes would be produced at intermediate speeds where $n$ is $\frac{1}{2}$, $1\frac{1}{2}$, $2\frac{1}{2}$, etc. The diagram pictures the situation where $n$ has the values 1, $1\frac{1}{2}$, 2, $2\frac{1}{2}$ and 3.

The boat must do work to produce these waves in the wake. You can see from the diagrams how it does this work. The bow of the ship is always in a wave crest of its own making. The water pressure against the bow is greater than if the wave were not there, so extra work must be done to move the boat forward. The stern tends to be in a trough of its own making and the water pressure on the stern is then less than if the trough were not there; this again requires extra work to be done if the boat is to move forward. If the speed is such that the two wakes tend to cancel, the trough around the stern is reduced and this reduces the force necessary to propel the boat. If the bow and stern wakes tend to reinforce and give a large wake, then the trough around the stern is exaggerated and the work needed to propel the boat becomes greater.

At the beginning of Chapter 5 I mentioned the large tanks and model boats that are used to test new designs for ships. People working with these models find that the force needed to tow a model varies in an oscillatory way with the speed at which it is being towed. The 'wave-making resistance' changes, as we have just seen.

## The wake of a fishing-line

Small objects produce wakes too, as you can see by moving a thin stick through still water. You can see the wake better if you can find a place where a stream is moving fairly quickly (about 1 foot or more per second). Then you hold the stick stationary in the water and the wake forms round it. If you have fished in quick-moving rivers you will have seen this, which is why I called this section "The wake of a fishing-line".

The wake of a fishing-line or a small stick does not show the arrow-head patterns that we found in the wake of a boat. This is because the small-scale waves travel by a different law. They are surface tension waves or 'ripples'.

I have sketched a fishing-line wake in the diagram. Notice that some of the wake waves are ahead of the fishing-line, upstream from it. Let's see if we can account for this.

You can think of the fishing-line as producing a succession of small disturbances in the water, and think of circular ripples spreading out from each one. So far we are arguing in the same way as we did with the wake of the bow of a boat, but now comes a difference. These spreading waves are ripples. At the end of Chapter 4 I found

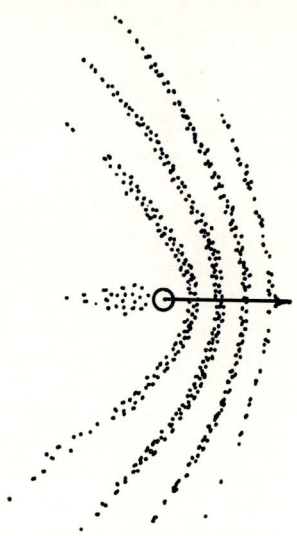

Fig. 7.10. The wake of a small obstacle (or fishing-line) in a stream is ahead of the obstacle.

how to include surface tension in the wave formula. The equation that I quoted was:
$$c^2 = g/k + Tk/\rho.$$
If we are thinking of very short ripples we could even forget the effect of gravity and write the formula just as:
$$c^2 = Tk/\rho.$$
We need to find the group velocity too. If the above formula is rewritten putting $2\pi/k$ for $\lambda$, like this:
$$c^2 = 2\pi T/\lambda\rho$$
one can differentiate it and get:
$$2c\, dc/d\lambda = -2\pi T/\lambda^2\, \rho$$
On rearrangement this gives:
$$\lambda\, dc/d\lambda = -2\pi T/2c\lambda\rho$$
$$= -c^2/2c = -\tfrac{1}{2}c$$
So according to the formula for group velocity, the group velocity of ripples is:
$$U = c - \lambda\, dc/d\lambda$$
$$= c + \tfrac{1}{2}c = 1\tfrac{1}{2}c$$
Apparently the group travels faster than the ripples themselves.

In thinking of an isolated group of ripples you need to imagine the ripples being continually left behind to fade away, while new ripples continually appear ahead of the group. This is the reverse of the behaviour of gravity waves.

But now we go on with the argument in much the same way as before. If the ripple has a radius $r$ at time $t$ after the start, it forms part of a group whose velocity is $r/t$. But the actual speed of the ripple, $dr/dt$ is only two-thirds of this:

$$dr/dt = 2r/3t.$$

You can perhaps show that this means that the radius varies according to the law:

$$r = At^{2/3}.$$

The number $A$ is not yet decided but it will be different for each ripple in the spreading pattern.

You can see from the diagram that a spreading ripple produces a 'wake' wave that is ahead of the fishing-line or stick. To make this diagram I took the point marked 0 to represent the position of

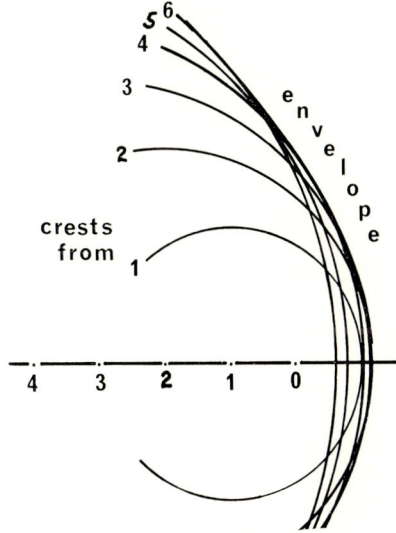

Fig. 7.11. Constructing the 'wake' of ripples from a fishing-line. Here I travel with the water and see the fishing-line moving to the right.

the stick and picked off a succession of equally spaced points on the path, and marked these by numbers 1, 2, 3, etc. These numbers are proportional to the times elapsed since the disturbance occurred there. I have drawn circles centring on these points, with radii proportional to the $\frac{2}{3}$ power of these times. In order to make a diagram I arbitrarily chose the circle centred on point 1 to have a radius of 2 units, and using the $\frac{2}{3}$ power law I found that the various circles should have radii:

$$2, \quad 3\cdot 18, \quad 4\cdot 16, \quad 5\cdot 04, \quad 5\cdot 82, \quad 6\cdot 60.$$

You can see that the envelope is a curved wave starting ahead of the stick. The physical reason is that a ripple begins by moving faster than the stick, so that it moves ahead of it. There comes a time when the speed of the ripple has decreased till it is just equal to the speed of the stick. The ripples from successive disturbances then agree in position and begin to reinforce and create the wake wave at some point ahead of the stick. When the speed of the ripple gets still less the reinforcement occurs not directly ahead but to either side and the envelope curves gently backward as you see.

Other ripples in the spreading pattern produce envelopes of other sizes but of the same shape and they all lie ahead of the stick. The ripples directly ahead are travelling at the speed of the stick so they are equally spaced by the wavelength that suits that particular speed. The quicker the speed, the smaller is this spacing. In fig. 7.10 I have sketched a number of envelopes to show what the ripple wake looks like.

We don't seem to have accounted for any waves directly in the rear of the stick, yet if you try the experiment I think that you may find some. This is because we have forgotten the gravity waves. There will be some spreading wave crests whose wavelengths are so long that gravity is more important than surface tension. These will accelerate, increase speed as they travel outward and create a wake in the rear as we saw when discussing the boat.

The combined action of gravity and surface tension leads to a minimum possible speed of waves on water. It works out at about 23 cm s$^{-1}$ for clean water. If the stick or fishing-line is moving relative to the water at less than this speed, then no wake waves are formed, neither ahead nor behind.

*The smooth track left by a ship*

If you have travelled on a large ship you may have noticed that the actual track along which the ship has passed is often distinguishable for a considerable distance astern. The track it leaves in the water is curiously smooth. There may be short choppy wind waves on the rest of the sea but the actual track of the ship is free of them; they seem unable to cross that region of water, though the longer waves on the sea do so quite easily.

I think that one reason for this is that the water in the track is in motion through the action of the screws that propel the ship. The purpose of a screw is to pick up water and send it backwards in a jet.

This drives the ship forward according to the momentum principle. But a rotating screw always tends to make the water rotate, too. Large ships usually have twin screws rotating in opposite senses. You can see two possibilities here. A ship designer always arranges that the screws shall be 'outward turning', that is to say the blades of the two screws as they reach their upper position are moving outwards away from the centre line of the ship†. So the water along the track of the ship is left moving backwards, of course, but it is also circulating, rising along the centre of the track, then passing outwards at the sea surface and sinking on either side as the diagram shows. The result is that when waves try to cross the track of the ship they have to fight their way against this opposing motion of the surface water. Long waves, which travel quickly, can do so. Short waves travel slowly and may be unable to do so, and this leaves the track smooth in comparison with the rest of the sea.

I wonder if this explanation is a good one. It is only my personal guess. I have some misgivings about it because one of my friends recalls crossing a large lake by ship and being able to see the track stretching back behind the ship to a place that had been passed 10 minutes before. I would scarcely have thought that the water would continue circulating for so long a time. Perhaps you yourself can think of it differently and argue about ship tracks in a different way. I leave it to you. Do you think it likely that the screws are spreading a film of oil on the water?

Yet it is certain that waves can be prevented from entering a region of moving water, and in the rest of this section I shall try to explain why this is so and point out some other consequences of the effect that moving water has on waves.

*Waves riding on a stream*

In the diagram I have pictured water moving steadily in a stream from the right where it is relatively shallow to the left where it is

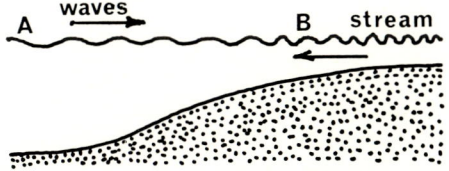

Fig. 7.12. Waves entering a stream.

deep. As the water advances, its speed will gradually grow less. In the very deep water on the left I shall suppose that the velocity is small enough to be ignored. What happens to waves that come from

† Because the boat steers better. I don't know why.

the left and try to enter this stream? It is simplest to think of a regular train of waves moving to the right past a point A in the deep water. What will the waves be like as they pass some point B where the water itself has a speed $V$ towards the left?

The waves that pass point A are all exactly alike so all will behave in the same way. There seems no reason for wave crests to disappear and I shall assume that none does. The alternative possibility is that they all in succession disappear at some point. Assuming that they have not disappeared at point B, I think you will agree that the same number of waves per minute passes B as passes A.

But the waves that pass B must look different from those that pass A. They must have a shorter wavelength. If they had the same wavelength at the two places they would travel through the water at the same speed†. But at point B the water is moving and the speed of the waves relative to the ground is less. To get the same number of waves passing per minute the wavelength must be less near point B.

But if the wavelength near B is smaller, then the speed through the water must be less and this means that the wavelength must be yet smaller. The argument is becoming complicated. You can use algebra to find where it leads.

*Finding a formula*

When one is expressing the problem mathematically it becomes easier to remember what the symbols mean if a positive velocity always means a velocity towards, say, the right, while negative velocity means towards the left. In the next diagram, fig. 7.13, I

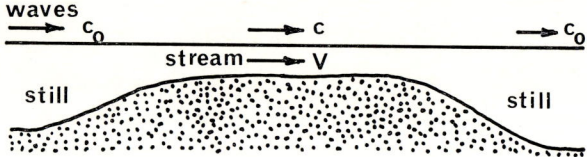

Fig. 7.13. Finding a formula for the way that waves on a stream will behave.

have redrawn the situation so that the water is moving to the right in the positive sense, and if the waves are moving also in that direction over the water then the velocity of the waves is positive too. I have also supposed that there is another deep-water region on the right where the speed of the water again becomes negligible. This diagram allows for many different possibilities. The waves may enter from either side and the water may flow either way.

† I am assuming that the water is everywhere deep enough for the waves to obey the deep-water formula. Then equal wavelengths mean equal speeds.

134

Suppose that waves are coming in from the still water on the left. I will write their speed as $c_0$ here, and it is positive. If the wavelength here is $\lambda_0$, then the number of waves passing a fixed point per unit time is just:

$$c_0/\lambda_0.$$

If the waves pass through the stream into the still water on the right, the speed and wavelength again become:

$$c_0 \text{ and } \lambda_0.$$

But what happens at some intermediate point where the water is moving with a speed $V$, which is towards the right if $V$ is positive? The speed of the waves relative to the water I will write as $c$, but their speed relative to the ground is $c+V$. So if the waves here have a wavelength $\lambda$ the number of waves per unit time that pass a fixed point is:

$$(c+V)/\lambda.$$

But we agreed that the same number of waves must pass every place. So we can write:

$$c_0/\lambda_0 = (c+V)/\lambda.$$

I am now going to suppose that the wave speed and the wavelength are connected by the formula for waves in deep water. This will simplify the argument and it might well be true. I spoke of the seabed as having different depths, but this was only to suggest an actual way in which the moving water might have different speeds at different places. It could well be that the water was still so deep (or the waves so short) that the waves progressed according to the deep-water formula:

$$\lambda_0 = 2\pi c_0^2/g, \quad \lambda = 2\pi c^2/g.$$

Then I could write:

$$1/c_0 = (c+V)/c^2.$$

This can be rewritten as:

$$c^2 - cc_0 - Vc_0 = 0.$$

It is a quadratic, and the solutions are:

$$c = \tfrac{1}{2}c_0[1 \pm \sqrt{(1+4V/c_0)}].$$

This allows us to calculate the speed of the waves at a place where the speed of the water is $V$, if the speed in still water is given.

But isn't it curious that the equation should be a quadratic and offer two answers? Surely in the problem as I have stated it there is only one possible speed that the waves can have at a particular place. The waves cannot have a choice of doing one of two things.

To make sense of this algebraic solution we had better draw a graph of the result, plotting the wave speed $c$ against the water speed $V$.

I have made this diagram and lettered some points on the curve, L, M, N, O, P.

For point M, the physical meaning is quite clear. Here the speed of the stream is zero ($V = 0$) while the waves have a speed $c_0$. So this just represents the wave-train on still water.

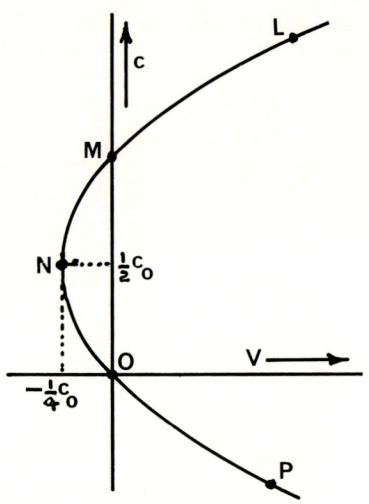

Fig. 7.14. A graph made from the formula.

If these waves move to a place where the stream velocity begins to be appreciable then the change in $c$ is to be found by moving a little way along the curve to the appropriate value of $V$. In our original problem $V$ was opposite to the wave travel, and therefore negative, so the part of the curve between M and N represents what happens when waves enter an opposing stream; the wave speed becomes smaller. On the other hand, if waves enter a stream travelling in the same direction, then $V$ is positive and we move along the curve from M towards L and beyond. The wave speed relative to the water increases; it also follows that the wavelength grows longer too.

You will notice that the branch of the curve from M to N does not go beyond N. Our formula gives no answer if the speed of the stream opposing the waves is greater than $\frac{1}{4}c_0$. Does this mean that waves cannot penetrate to those parts of the stream where its speed exceeds $\frac{1}{4}c_0$? I think that it does and in fig. 7.15 I have pictured what I think would happen. The physical reason is as follows. At point N the speed of the waves has been reduced to $\frac{1}{2}c_0$, which is still twice the speed of the stream, so the wave crests are still moving upstream. But in deep water the group velocity is half the wave speed. At point N the group velocity is $\frac{1}{4}c_0$, just equal to the speed of the stream.

Since the two are in opposite directions the group can advance no farther upstream. This seems to me to mean that if a regular succession of waves come upstream from the still water (point M), then they will successively reach point N and then still moving upstream will pass beyond the group and one by one fade out.

You may well ask what becomes of the wave energy in this case. Waves are continually bringing energy into the stream, yet all the waves finally fade out. Where has their energy gone? It seems to be carried as far as point N and left there, because the groups can advance no farther. I think you will agree that as the groups advance upstream and move more slowly they must crowd together, making the energy per metre of sea surface greater as they do so. The waves will then be higher. I think that as the waves move upstream towards the limiting point represented by N on the curve, they will increase in height till they foam at the crests. This dissipates the wave energy. Then one by one the waves would fade out as they pass the limiting point N.

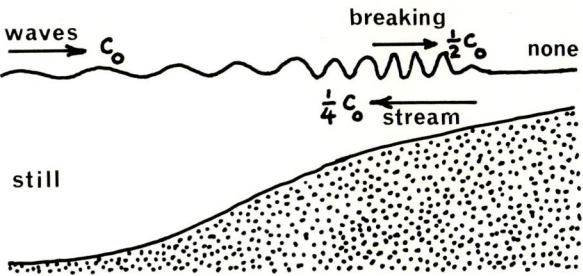

Fig. 7.15. Waves are blocked if they enter an opposing stream whose speed exceeds one-quarter of that of the waves when they were on still water.

So it seems that no wave energy can penetrate against an opposing stream whose speed is as big as one-quarter the speed of the waves on still water. I think that this explains the smoothness of the ship's wake. If the screws left the water flowing outward at only 25 centimetre per second this would still prevent the passage of waves whose normal speed would be 1 metre per second. You can see from this formula:

$$\lambda = 2\pi c^2/g$$

that if $c$ is 1 metre per second then $\lambda$ is 64 centimetre:

$$\lambda = 6 \cdot 28/9 \cdot 8 = 0 \cdot 64 \text{ metre}.$$

Waves of this length or less would be kept out of the track of the ship.

*The wind and the tide*

Seamen have a saying that waves are likely to be dangerous where wind and tide are opposed. One can see that when water moves in the opposite direction to the wind, the relative speed of wind over the water will be slightly increased. I do not think, however, that this is a sufficient reason for the waves being very steep and breaking. I think the reason is that tidal streams are enhanced in local regions, at estuaries or off headlands. In the rest of the ocean the wind will generate waves in the normal way. I think that it is when these waves pass into the region of a tidal stream moving in the opposite direction that they become both shorter and higher and so more dangerous to small craft.

*Other waves on a stream*

We have not yet thought about the branch of the curve that I labelled N, O, P. What kind of waves are represented here? If you look at the part NO you will see that the waves and the stream move in opposite directions ($c$ positive, $V$ negative). Also $V$ is numerically less than $c$ but greater than $\frac{1}{2}c$. So the waves are moving upstream but their energy is being carried downstream. You could imagine such waves as having been created on the stream itself by wind, or by a moving boat as I have suggested in the diagram. There are two

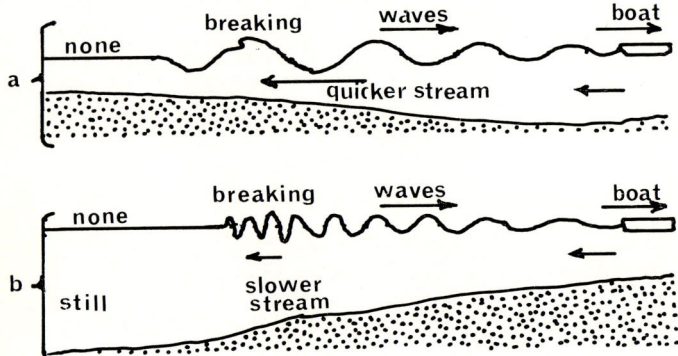

Fig. 7.16. Two curious things that should happen but which I have never noticed.

possibilities; in the first, the energy may be carried to places where the speed of the stream is greater. The waves there will be longer and faster and there comes a point corresponding to N on the graph at which the group velocity is just equal (and opposite) to the velocity of the stream. Energy will accumulate here and the waves will be foaming at the tops. Yet oddly enough it would seem that wave crests should continually emerge from this region of breakers and move upstream against the current.

The other possibility is that the energy may be carried towards a region where the stream velocity is less, or even zero. Then one moves along the graph towards point O. The waves become shorter, and they become steeper too because the group velocity is growing smaller and the energy accumulates. Here again there would be a region of foaming waves with wave crests emerging from it to move upstream.

I have never seen either of these things happen. Perhaps if I worked in a hydraulics laboratory I would arrange an experiment to test the ideas. Perhaps you may see them happen if you spend some time near a river.

The branch of the curve labelled OP represents rather similar situations, in that the waves and the stream have opposite velocities. But now the waves are so short that the wave crests themselves are carried backwards by the stream. These waves would break if the stream carried them into still water (point O), but they could be carried to places where the stream velocity was greater and this would make them longer and lower; they would not break there.

So it seems that the branch N, O, P represents waves that could never emerge into still water, while the branch L, M, N, represents waves that can.

*Goodbye*

Now I must stop. Waves can do curious things, not all of which are understood. And there are many kinds of waves besides water waves. If you go wave-hunting, may you have good luck.

*A postscript*

Two people wrote this book. You can send a message to them if you wish, perhaps to point out a mistake, or perhaps to talk about an idea. Post your letter to the editor :

> The Editor,
> Wykeham Publications (London) Ltd.,
> 10–14 Macklin Street,
> London, W.C.2.

and start it like this

*Dear Sir,*

*Please forward this letter to the people who wrote the book called " Water Waves " in 1969.*

Then go on to say what you wish. You will get a reply if you remember to put your address after your name. They would want to reply to you.

# INDEX

Acceleration
— judged from surface tilt   7
— pattern in waves   11
Arrow-head waves   125, 126

Barbados surf   80
Beat of the surf   108
Bernoulli, D.   30
Bessel functions   70–74
Binary counting   90
Bob and curtsey   62

Cartesian coordinates   33
Circular waves   70–74
— —, manner of spreading   121
Circulation   22–29
— theorem   23
Conservative field of force   29
Corner-to-corner waves   68
Correlation coefficient   110–114
Correlogram   114

Deep-water waves   18
Differentiable function   24
Disappearing waves   77

Earthquakes   17
Energy in waves   90
Envelope of waves   123, 124
Exponential functions   40

False vertical   5
Flow   23
—, changes in   24–28
Free surface, constant pressure   6–7, 47, 49
Frequency analyser   84
— spectrum of waves   85

Great circle path of swell   86
Group behaviour   76
— velocity   78–80

Height of waves at sea   93
High waves   92
Huygens' construction   123

Ideal fluid   21
— wave-train   9
Idealization in physics   8–9
Internal waves   94
Irrotational motion   30

Knot (sea miles per hour)   82

Lag correlogram   114
Long-crested waves   10

Microseisms, detection   105
— from waves   103–105
Model ships   76
Motion below surface   19, 48, 50, 52–54, 56–57
— of ideal fluid   20

One-third highest wave   93

Paraffin wax for models   76
Partial differentiation   35
Particle motion in wave   50–57
Pendulum   45
Potential energy in wave   90
Power from waves   91
Pressure, free surface   47, 49, 53
— in steady flow   43
Profile of ideal wave   93
— — — —, formula   48

Raft in waves   8
Rays   99
Refraction of waves   97–101
Rip currents in surf   107
Ripples   74

Sea miles   82
Seiches   65
Separation of variables   39
Shallow-water waves   15
Shore, waves approaching   96
Short-crested waves   63–64
Speed of ripples   75
— — waves   13, 49
Standing waves   59, 61, 64–65

'Stationary' and 'standing'   60
Steady state   36
Stevenson, R. L.   93
Streams and waves   87, 133–138
Submarine   51, 96
Surf at Barbados   80
—, beat of   108
—, rip currents in   107
—, swimming through   1
Surface tension waves   75
Surges from shore   109
Swell
— from distant storms   83, 84, 89
—, coming ashore   97
—, recording of   83
—, time of travel   90

Tanh function   53
Tidal streams, effect on wave period   87
Tides as a wave   15
Tilted surface   3
True vertical   4
Tsunamis   16

Velocity
—, group   78–80
—, particle   11, 24–26, 46, 48, 52

Velocity potential   31
— —, deep-water wave   42
— —, radial collapse   32
— —, steady stream   37
—, ripples   75
—, waves   15, 49, 53, 54
Vertical
—, false   5
—, true   4
Volume change, rule for zero   33

Wake of fishing line   129–132
— — ship   118–128
— — —, interference of   128
— — —, shallow water   127
Wave groups   76–77, 120
— rays   99
Waves
—, coming ashore   96
—, deep water   18, 42–49, 56
—, not the water   2
— on fluctuating stream   87
— — steady stream   133
— over a bed   51, 57
Weather map, synoptic meteorological map   81

Zero change in volume, rule   33–35

# THE WYKEHAM SCIENCE SERIES
*for schools and universities*

1  *Elementary Science of Metals*    J. W. MARTIN AND R. A. HULL
   (S.B. No. 85109 010 9)*

2  *Neutron Physics*    G. E. BACON AND G. R. NOAKES
   (S.B. No. 85109 020 6)*

3  *Essentials of Meteorology*    D. H. McINTOSH, A. S. THOM
   (S.B. No. 85109 040 0)*    AND V. T. SAUNDERS

4  *Nuclear Fusion*    H. R. HULME AND A. McB. COLLIEU
   (S.B. No. 85109 050 8)*

5  *Water Waves*    N. F. BARBER AND G. GHEY
   (S.B. No. 85109 060 5)*

6  *Gravity and the Earth*    A. H. COOK AND V. T. SAUNDERS
   (S.B. No. 85109 070 2)*

7  *Relativity and High Energy Physics*    W. G. V. ROSSER
   (S.B. No. 85109 080 X)*    AND R. K. McCULLOCH

Price per book for the Science Series **20s.—£1.00 net** *in U.K. only*

# THE WYKEHAM TECHNOLOGICAL SERIES
*for universities and institutes of technology*

1  *Frequency Conversion*    J. THOMSON,
   (S.B. No. 85109 030 3)*    W. E. TURK AND M. BEESLEY

Price per book for the Technological Series **25s.—£1.25 net** *in U.K. only*

\* Standard Book Catalogue Reference Number.